Global Claims in Construction

Ali D. Haidar

Global Claims
in Construction

 Springer

Dr. Ali D. Haidar
Dar Al Riyadh
PO Box 271055
Riyadh 11352
Saudi Arabia
e-mail: alidhaidar@aol.com

ISBN 978-0-85729-729-7   e-ISBN 978-0-85729-730-3
DOI 10.1007/978-0-85729-730-3
Springer London Dordrecht Heidelberg New York

British Library Cataloguing in Publication Data
A catalogue record for this book is available from the British Library

Cover design: eStudio Calamar S.L.

Printed on acid-free paper

Springer is part of Springer Science+Business Media (www.springer.com)

To my wife
Joud and Neil

Preface

For anyone embarking on writing a global claim and for professionals charged with the responsibility of determining the rights and liabilities of the parties with regard to delays, disruptions, total cost method, modified total cost method and global claims, this book is an essential reading.

Global claims are unique as they involve combining all the delays and disruptions caused in one project under one heading and putting the claim forward to the client. They assume that all the problems encountered on site cannot be separated, and therefore the losses incurred are initiated by the client and his representatives.

Historically, the courts had been reluctant to accept global claims. Recently, the situation has been shifting and the courts have been accepting these types of claims or a modified version of them. This is due to the distinctive characteristics of the newly constructed projects and the types of problems envisaged on the very large projects that are common nowadays. The complexity of the activities performed on a construction site, the great number of professionals and entities that are involved in the execution, the new types of contracts involving design, procurement and build and the procurement process, where much of the materials are supplied from different countries, make the process of individual claims, where each cause has a distinctive effect and a distinctive loss, almost impossible.

Even if the process of identifying individual causes and their effects and establishing the specific damages can be achieved, the time that can take to present individual claims is prohibitive especially if the task could be repeated hundreds of times on the very large projects. The courts are also showing reluctance to screen thousands and thousands of documents just to prove liabilities or to decide on variation issues.

Therefore, global claims are looking more and more the way forward of writing and presenting claims, and the future will see more hybrid species of claims combining global claims and standard claims.

Contents

Chapter 1
Introduction

1.1 Synopsis

This book is intended to give an overview of the nature of the law as it applies to construction. It provides the general principles to a particular context that should assist in getting grips of understanding the basic rules in contractual disputes, damages, how and where to find legal rules and mostly how to apply rules in the construction industry. It will also provide directions when the rules are broke or where non-existent in order to provide basis for claims.

When establishing a claim, it is sometimes essential to consider matters liable to defeat the contract by rendering it void, voidable or unenforceable. These include mistake, misrepresentation, duress, undue influence or special principles of frustration. Principles on privity, contract proferentem and the legality of letters of intent are included as well as they are imperative in understanding the special doctrines applicable in construction contract law. In most cases, the contract is stated as such where the injured party cannot find remedies within the scope of the agreement signed. In other instances, it can be possible to find relief within certain clauses and then seek further remedies within vitiating factors such as duress or mistake. A global claim can then be established based on the fact that all these matters are intertwined as such the claimant cannot find the relationship between the damages incurred and the related losses.

This book provides an explanation of delays, disruptions and damages, which is presented in a straightforward accessible manner. The book also identifies time-related issues and the main effect of an extension of time, not as often incorrectly believed to give rise to an automatic entitlement to compensation for prolongation costs, but to relieve a contractor of its liability for liquidated damages during the period of the extension.

The subject of damages is covered in detail including the calculation of the different categories such as direct and indirect costs, equipment, materials, labour and others. Likewise, the damages suffered by the client, either liquidated or penalties, are also explained.

A. D. Haidar, *Global Claims in Construction*,
DOI: 10.1007/978-0-85729-730-3_1, © Springer-Verlag London Limited 2011

The book has been written as a code of good practice to be used to present global claims during the administration of the contract, including the assessment of total cost claims and to establish a mechanism for quick and fair settlement. The book will recommend that a proper calculation methodology should be submitted by the contractor and applied by using the bill of quantity as basis for the calculation methodology being the total cost or the modified total cost method. The calculation should show the manner and sequence in which the contractor has suffered losses to carry out the works and the mitigation steps he has undertaken.

In this book, global claims are described comprehensively and a detailed case study of a real example is presented on how to write a global claim using the modified total cost method analogy. This case study will be a valuable tool for professionals embarking on writing claims and particularly global claims as it shows the structure of thoughts the claimant must apply to the matrix of facts and the complex data in order to present his global claim in a proven manner.

Construction claims have a language and procedures of their own. A party to a construction dispute must elect between inconsistent claims, defences and remedies. An uninformed choice or failure to pursue a particular remedy or a particular defence can be a costly mistake. This book highlights the issues related to global claims which are available. There will be also much emphasis on case law to show how the courts have looked and analysed issues raised in this book.

In principle, this book intends to:

1. Demonstrate a basic understanding of the nature, operation and principles of the law of contract. This book will emphasise in this part mainly on English law as it is, historically, the most established in relation to this subject.
2. Identify the meanings of offer, acceptance, consideration and intention to create legal relationships.
3. Demonstrate a basic understanding of how the interpretation of contractual terms affects the enforcement of the agreement.
4. Identify the factors that lead to the termination of a contract and the practical consequences and remedies which may then arise.
5. Familiarise the parties as to the contractor and client duties, their agent's responsibilities and how the parties must act in a fair and reasonable manner to avoid disputes.
6. Identify disputes that may arise and lead to claims and mitigation measures.
7. Demonstrate a basic understanding of the measure of damages in relation to breach of contract.
8. Identify damages and calculation of losses.
9. Describe global claims, total cost methods and modified total cost methods in depth. A case law approach to look on how the courts approach these types of claims.
10. Calculate losses using the total cost and the modified total cost method.
11. Finally, a case study, on how to prepare a global claim with its detailed calculation analysis, is presented to assist professionals in writing and defending global claims generally.

In summary, the book aims at being a useful tool for professionals whose work involves them in the construction industry, including public and private clients, construction managers, main contractors, subcontractors, designers and lawyers. The audience intended to read this book include also engineering and technical managers, quantity surveyors, company directors and students in engineering, construction law and management.

1.2 Importance of Contract Law for Disputes

The idea that contractual obligations are based on agreement must be qualified in relation to the scope of the principle of freedom of contract. In the nineteenth century, judges took the view that persons of full capacity should in general be allowed to make what contracts they like. The law only interfered on fairly specific grounds such as misrepresentation, undue influence or illegality.

The first requisite of a contract is that the parties should have reached agreement. Generally speaking, an agreement is made when one party accepting an offer made by the other. Further requirements are that the agreement must be certain and final. An offer is an expression of willingness to contract on specified terms, made with the intention that it is to become binding as soon as it is accepted by the person to whom it is addressed. An offer may be addressed to an individual, or to a group of persons, or to the world at large; and it may be made expressly or by conduct. Acceptance is a final and unqualified expression of assent to the terms of an offer.

The objective test of agreement applies to an acceptance no less than to an offer and should contain the following elements:

1. Continuing negotiations.
2. Acceptance by conduct.
3. Acceptance must be unqualified.
4. The battle of forms.
5. Acceptance of tenders.

In law generally, a promise is not, as general rule, binding as a contract unless it is either made in a deed or supported by some consideration. A contract may be rescinded or varied by subsequent agreement. The object of rescission is to release the parties from the contract, while that of variation is to alter some term of the contract.

Contracts need to be understood within the framework, and against the background, of common law and statutory doctrines. The significance of such doctrines varies. Some doctrines empower the contract by giving it legal force, while others allow the contract to be treated as terminated or avoided in certain situations. Doctrines introduce terms into the contract, impose liabilities and, in some instances, allow entitlements outside of a contract. Some common law and statutory doctrines apply as default positions; they apply if no specific provision is made or unless they are excluded by express words. Some apply irrespective of the words of the contract. Others are, apparently, applied if the words of the contract yield too harsh a result.

Doctrines in law are also relevant in that some construction contract provisions are drafted to emulate them. For example, financial entitlements are framed in terms of *quantum meruit* (payment for work) and damages (equivalent to breaches of contract) situations. This assists in understanding and interpreting the contract provisions. It also allows equivalent remedies or relief to be granted within a binding contract, side stepping the undesirable consequences of the contract being rescinded or terminated through common law processes.

Some vitiating factors in law such as misrepresentation, mistake, negligent statement, foreseeability and frustration are often adopted to relieve contractors (and clients) from their obligations and to provide them with remedies when the contracts they have entered into fail to do so. The parties to a contract, when subjected to these factors, could rescind or repudiate the contract.

1.3 Construction Characteristics: Contracts and Relationships

Whatever contractual documents are used, there are a number of characteristics that are common to, and largely distinctive of, almost all construction projects. It is generally in these characteristics that the seeds of eventual disputes lie, and which explain the higher occurrence of disputes in the industry and the complexity of identifying and preparing claims to calculate the damages caused by these disputes.

The prototypical characteristics of construction accentuate the significance of responsibility for design and procurement. The high degree of inter-activity in the design and construction process, between client, his appointed consultants and agents, and the contracting team distinguishes construction from many other industries. However, in some new large housing projects and the mega projects with repetitive types of structures, there is a tendency of automation but still factors such as the soil, type of construction, precast as opposed to cast-in concrete or steel structure and the type of finishes will make each project unique.

Though the major contracting entities of today have moved far from what was envisaged as a classic design and build scheme, the structural arrangements of the construction industry have remained largely unchanged, save perhaps for more recent initiatives in the provision of privately financed infrastructure works.

What contract may be used, the actual works on almost all construction projects exhibits recurrent distinctive characteristics:

- The prototypical nature of the works.
- Split responsibility for specification and design.
- High degree of inter-activity between client, contractor and supplier.
- Expectation of, and provision for, substantial levels of change to the specified.
- Scope of work, complexity of sequencing of activities and dependencies on other activities/supplies.

- Site specificity.
- Exposure to, and dependence on, outside factors such as international fluctuations in raw materials prices, lack of resources due to demand and acts of god.
- Longevity of the products, and lateness of revelation of defects.
- The diversity and sheer volume of evidentiary material.

The delivery of a product in construction is a process, not an event. The process requires many participating entries. Even where the main contractor takes on the design responsibility, the client is still likely to have a consultant team. Also, key elements of supply, such as plant, machinery and materials, especially long lead items, may have to be sourced from a specified subcontractor or supplier.

Typically, in practice, the participants in a construction project will include the client organisation, advised by architects, engineers from various disciplines and costs consultants, a main contractor usually with sub-consultants, and a host of sub-contractors, labour supplying agencies and materials suppliers.

Traditionally, employers have engaged professionals to manage construction and engineering contracts. While the scope of their duties depends on the terms of the particular contract, usually they perform two distinct roles. The first is as an agent of the employer in issuing instructions and ordering variations. The second is as a decision-maker in certifying payments, assessing claims for loss and expense and in awarding extensions of time.

An engineer or a contract administrator having the role as a decision-maker has to act independently, impartially, honestly and fairly. He must not favour either contractor or employer. However, he does not have to apply the rules of natural justice when making his decisions. If the administrator negligently over certifies in the contractor's favour he can be held liable to the employer who engages him. In special circumstances, he might also be liable to third parties such as institutions lending money to the employer. In normal circumstances, an engineer who under certifies will not be liable to the contractor.

However if, for example, he makes gratuitous representations to the contractor he may be found to have assumed a responsibility to him and be liable in negligence. If the employer exerts pressure on the administrator so that he loses his impartiality and independence then the administrator's certificate may be invalid and his decision ignored. Furthermore, if the employer knows that the engineer is not carrying out his functions properly then he may himself be liable to the contractor for breach of contract if he does not take steps to correct the position.

1.4   Costing in Construction

In most cases, a contract will be awarded after a competitive bidding process, whereby the contractor awarded the job will be the one who gave the lowest bid for the work. This will generally be the result of extremely optimistic calculations that have been made by the contractor, calculations that have allowed the contractor to arrive at a price for the job that is lower than more moderate or conservative calculations made by the unsuccessful bidders.

Any contractor, who factors in all of the risks and makes contingencies for some of the uncertainties facing the project, will rarely come in at the lowest bid. In fact, it is reasonable to conclude that in the case of the lowest bid winning a contract, it is not the contractor that is best placed to perform the task on time and on budget who will be awarded the contract, but rather, it will be the contractor most desperate for the work, and therefore most willing to make the unrealistic promise without planning for difficulties that could arise. Herewith, foster most of the issues in claims where they are brewing even before the contractor steps on site.

A rational system for the determination of the price for construction works might involve three principal elements:

1. The tendered price for which the contractor is willing to do the work;
2. Some method of assessing the suitability of that price, by way of a breakdown; and
3. A method or schedule for pricing any additional works or changes to be made to the scope within the tendered price.

There are also structural elements in the arrangements for the construction industry which exacerbate the uncertainty of the sums eventually due. Fixed prices (or lump sums) are not the normal pricing method in many projects. On some projects, works of uncertain extent are commissioned to contractors working on special rates; and arguments will arise as to the efficiency of the works carried out and the accuracy of the resources recorded to justify the rates claimed. Major projects for civil engineering works have traditionally been so uncertain in their scope, by definition, that the new contracts are drafted on the basis that they will not be tendered for a fixed price. Rather, the payment system is based on a remeasurement basis. Some even are procured on a cost plus basis.

Thereafter, the timing of payment of the price would be a matter for negotiation upon a number of factors, with each side looking for a form of security as working in progress is delivered into capital value. Any such system is likely to involve payment in stages, against identified milestones including perhaps some retention to a final stage for the assessment of outstanding works or defects. In many industries, it is the suppliers that have the special expertise in the design and construction and to ensure that their capitalisation is strong enough to secure, in large measure, the financing of their production until their products are delivered to their customers.

Furthermore, the traditional payment systems have the characteristic of postponing the resolution of uncertainties until late stages, or even after completion, of the project. Meanwhile, the financing of the contractors work is achieved by interim payments assessed by judgment evaluation by one of his representatives, being the architect, engineer or others. Every such judgment is capable of argument, contradiction, claim and eventual dispute.

1.5 Time in Construction

Programming of the complex sequences of activities and their dependencies is of course one of the principal skills of the successful contractor, or the construction manager. However, all but the most simple of projects will proceed from some such a programme; that is unmistakable in the construction industry, yet the traditional forms of contracts make little contractual provision to integrate the programming of activities into structural obligations. It is not an unfair generality to say that the contractor time-based obligation under the traditional forms comprises only an obligation to complete the totality of the works by a particular date.

The main parties involved in the contract are often reluctant to allow the initial tender programme to have the status of a contract document. This reluctance often stems from the traditional hierarchy of the construction industry, and the in-built fear that the parties involved lack the expertise in contractual techniques and furthermore the practical application of such techniques in order to achieve the target time and eventually cost.

As a result, the opportunity to make use of the contract and to identify early on the delays and disruptions to analyse existing disputes arising during the progress of the works and the more important post contract time-related disputes are lost. Disputes that ensue, quite often unnecessarily, leave the contracting parties with no means to reliably analyse the time and losses related damages.

Time-scaled networks of the as-planned, as-built and as-adjusted schedules are subsequently used to present the facts during claims negotiation, arbitration or litigation processes and are compiled using pain staking research of the site labour, materials and other records as required to produce the global claim quantification.

Other important issues arise in case of acceleration. When delays occur and where the client resists to grant the contractor an extension of time that he is entitled for an excusable event or where an extension of time has been granted by the owner, but for a shorter period than the contractor is entitled to, a contractor may feel compelled to accelerate the works in order to overrun the completion date set by the client thereby avoiding exposure to liquidated damages. This creates an intricate situation on the duty of care imposed on the client and his representatives and how the courts review such cases in providing relief for the contractor.

1.6 Changes and Variations in Construction

Changes in the delivered scope of a construction contract, or in the manner or sequence in which they are carried out, must to some degree be inevitable. This is the result of designing and constructing a complex prototype on a particular site, where its function includes interaction with external environment. Change orders and variations in fact arise for many reasons other than technical problems in fulfilling the contemplated design.

The freedom to make changes, after the contract has been let, has become a hallmark of traditional construction practice. The architect or engineer is given the power to order change that is often exercised. Its impact on the time for completion of the works, and their outturn cost can be greatly disproportionate to the extent of any particular change. The cumulative effect of such instructed changes can undermine the whole economy of a project if not well managed.

When ordered to carry out a variation, the contractor has the following options to choose:

1. The contractor will be entitled to refuse to carry out the instruction.
2. If the request is purportedly given under the variation clause and the contractor complies without objection, he will be estopped from subsequently claiming any right to payment other than under the terms of the contract.
3. If the request is made without any reference to the variation clause then the contractor will be entitled to a reasonable remuneration on the basis of an implied reasonable price under a separate contract.
4. On the strict basis of interpretation applied to exclusion clauses, the courts will construe restrictive provisions. For example, those requiring written instructions as a condition precedent, therefore, enable the contractor to recover a reasonable price for work outside its scope free of any such restrictions.
5. If work outside the scope is carried out, then the original pricing mechanism should only be departed from with respect to the work outside the scope and not any part of the original contract works.

Facts, when married to law, must persuasively demonstrate the desired and sought for result by virtue of the justice, equity and fairness of each party position. Therefore, complete documentation of change orders and delays during the life of the project is critical, particularly when most owners take the position that the total burden of proof is upon the contractor. Such proof, in order to be given any consideration, must be accurate, logical, valid by detailed legal and contractual analysis, and completely quantitative.

Change orders and variations must also be well managed in any construction contract and to include:

1. Clarification of time and cost risks to be borne by the principal.
2. Increased flexibility to vary the way the work is managed.
3. Detailed specification for the provision of programmes, method statements and historical progress records.
4. An ability to vary the resources, method or sequence of the works.

1.7 Documentation and Records

Masses of record materials are produced even on relatively small construction projects, many of them as crucial as they are informal. Project records may be as diverse as site investigation reports, feasibility studies, specifications, drawings, tender submissions, estimating and pricing details, minutes of meetings, formal instructions, test data, payment applications and certificates, weather records and so on. To all of that is added the great chains of correspondence between the participants, management reports in each of the entities, and the usual periphery of any business activity. It is haphazardly in these various forms of contemporary record that are to be found the clues as to the causes of disputed matters.

The primary objective of the contractor at the outset of the construction contract should be to ensure that the appropriate and necessary procedures, records, documentation and correspondence are established and maintained throughout the entirety of the contract so as to ultimately facilitate successful completion of the contract and to avoid delay and disruption and other claims ending in dispute. An equally important objective is to ensure that high standards of record keeping and documentation are maintained throughout the period of the contract to record the effect of delays, variations and other events and those procedures in respect of same are established and fully communicated to each of the contractor's relevant personnel involved in the contract.

Many delay and disruption disputes could be avoided if the parties properly monitored and recorded the above-mentioned information. Experts who advise on disputes often find that there is a lack of records resulting in uncertainty as to when a delay occurred, who caused the delay and the effects of that delay. Good record keeping can remove this uncertainty.

Records and information most likely to be crucial in the success of claims include:

1. Master programme identifying the critical path and indicating how the contractor had envisaged the sequence and timing of the various activities based on the tender information.
2. Progress schedule of activities against the master programme.
3. Estimate of weekly resources and anticipated expenditure to comply with the master programme.
4. Records of actual resources and expenditure based on progress.
5. Records of plant standing or uneconomically employed.
6. Labour allocation sheets and associated costs.
7. Variations register.

The reality is that a small proportion of time, money and effort expended by the contractor in putting in place the aforementioned procedures and record keeping at the outset of a contract could ultimately save him a significant amount of time, effort and money at the end of the contract in having to recover, in arbitration or court proceedings, loss and expense incurred due to delay and disruption to the contract. The burden and standard of proof that arbitrators and judges expect contractors to attain in order to substantiate their delay and disruption claims, is not attained by the majority of contractors who fall significantly short of this burden of proof and the standards of evidence required.

The recurring reason for this failure arises because the contractor has not put in place procedures, record keeping and the standards necessary to avoid claims becoming disputes. In summary, good records and procedures avoid disputes and, at the very least, ensure that if a claim does become a dispute, the contractor has the evidence to prove his claim.

1.8 Claims: Techniques and Structures

Delays and disruptions in construction projects are frequently expensive, since there is usually a construction leverage involved which charges interest, bonds and guarantees with ongoing fees related to the down payment and performance, management staff dedicated to the project whose costs are time dependent and ongoing inflation in wage and material prices. Equipment, labour and resources are usually tied to a project for certain duration according to its completion date, and when a project is delayed these resources can hinder the performance of other projects where they are due to be allocated.

Many techniques are applied to analyse delays and disruptions. Some of these methods have inherent weaknesses and should be avoided. This book points out the shortcomings of these faulty methods and explains how an analytical and a quantitative analysis should be performed. The three principal evidential aspects to proving delay and disruption claims are as follows:

- The requirement to prove that the event occurred and the extent of the event and its consequential effects. This is usually, but potentially not in every case, a question of fact, and is largely dependent upon the records of events as they occurred.
- The requirement to prove that the contractor has some entitlement under the contract or that the employer is otherwise liable for the delay event claimed. This is largely a question of law.
- The requirement to establish that the event had an effect on the completion date of the works. This will often require some sort of delay of analysis of the activities involved and proper records of the data that lead to this knock on effect.

A claim should clearly set out the following in order to be well structured and presenting the *causal nexus* that have applied in order to justify the justness of the methodologies presented:

- Factual events that have led to a delay or a financial loss and make a statement as to why the employer is responsible for these events.
- To show that each of these events is covered by the contract, and that any stipulations for making the claim that are contained in the contract have been complied with, such as time limits for making a claim.
- To show a strong causal link between the events complained of and the delay or financial loss. If there are a number of events leading to loss, then ideally, each

event will be apportioned its share of the loss that it has caused, and this will be backed up by contemporary documentary evidence.

Any notice required to be served under the contract should make particular reference to the relevant clause number under which it is being served and the contractor should follow precisely any procedure pertaining to the form or content of the notice. As notices generally require action by other parties, it is advisable, if not imperative, to keep a log of all notices given and of the dates by which response should be made by the employer (or its agent). It is infinitely better to point out to the other party, by way of a further notice, that there has been some inaction on their part, rather than keeping silent. There is often reluctance by contractors to serve notices with regard to delay or of a pending financial claim. Generally, however, the contract will contain a clear obligation on the contractor to provide these notifications. In serving such notices the contractor is merely exercising a contractual duty which is often a pre-requisite for safeguarding its own position.

The archetypal disputes in the construction industry arise of course from a contractor's claim to be paid more for the increased time and cost of additional works; and, the claim of building owners in respect of defects. The former is likely to be multi-party disputes as sub-contractor interests and responsibilities have to be taken into account. The latter, especially, may involve many parties; responsibility is sought to be attributed between designers, suppliers and contractors.

In construction litigation, claimants are increasingly seeking more exotic species of damages such as those flowing from constructive acceleration and other impacts to their work. Proof of direct causation between an act or omission, and the damages purportedly suffered is often tenuous or entirely incapable of proof. To overcome this impediment, contractors are turning with greater frequency to global claims and the total cost method of proving these damages. These types of claims are the main issues that this book will focus on.

1.9 Global Claims

Most people involved in commercial matters in the construction industry will understand what is meant by the term global claim. A global claim is a delay and disruption claim disguised. The difference between a global claim and a delay and disruption claim is that a global claim is a claim made for costs incurred due to delays and disruptions as a result of multiple events whose consequences have a complex interaction that renders specific relation between the event and the time or money consequence impossible or impracticable.

In its simplest form, a global claim allows a claimant to establish damages by calculating the difference between its actual costs of performance and the contract priced amount. This method can be of great assistance to contractors because in complex construction matters it is often difficult to quantify the actual damages resulting from a delay or disruption.

Claims, in their general form, should identify separately each item of claim together with its effect. Due to the difficulty in the fact that contractors and subcontractors often experience in providing this type of detailed information they often produce a global or rolled-up claim. This is one where no individual connection is made between each item of claim and its effect. Often the individual items of claim are identified but the loss is expressed as a composite whole.

Thus in a global claim manifestation:

1. A claimant can recover even though he is not in a position to demonstrate that any particular breach caused the loss; and
2. The wrongdoer rather than the innocent party bears the consequence of the impossibility of segregating the precise consequences of breach; and
3. If the breach materially contributes to the loss then, in principle, full recovery should be achieved.

The courts are increasingly demanding clearer explanations of cause and effect and in complex construction projects detailed time impact analysis. However, as this book will show that when it is impossible to separate the cause from the effect and when certain conditions apply the courts will consider favourably global claims and even when they fail to accept a global claim entirely they will consider apportionment of losses when the causes are split between the client and the contractor.

Global claims look sometimes as a simple way to present a claim by a lax contractor; on the contrary they are very complex claims and to be able to write a properly drafted global claim, the team in charge of writing a strong claim must be able to deal with the complex reasoning behind the claim and to understand that any mistake in writing the claim, even minor, can jeopardise the whole claim.

In order to prove a global claim, it is necessary for the claimant to satisfy the requirements of the particular contract under which he is operating. This is likely to require that a particular event is of a type that gives rise to an entitlement under the contract for an extension of time, and that the event has occurred and its extent cannot be quantified. In the context of construction projects, this is however often very much the case, and the extent to which this is possible is usually a function of the quantity and quality of the records held by the contractor about what actually happened on site.

A global claim in some instances is based due to disruptions not resulting in specific delays. Disruption has a different meaning to delay. In the context of a construction contract, disruption is the loss of productivity, disturbance, hindrance or interruption to the progress of a contractor. Unlike delay, the disruption may not lead to late completion of the work. Generally, the contractor is only able to recover disruption compensation to the extent that the employer causes the disruption. Although many contracts do not deal expressly with disruption, a contractor should maintain good site records to assist the client to make proper assessment of the disruption caused.

Even when a contractor cannot identify clearly the causes that ensued into damages or when he cannot establish the *causal nexus* between the disruptive events and the losses, he must still show that he has done all reasonable care to properly keep tracks and records of all the facts and the data related to this specific project.

It is intended that through the use of such claims the parties are encouraged to better articulate their respective cases and to exchange information and dialogue to enable them to better understand each other's case at an earlier stage and resolve their differences before trial.

1.10   Total Cost Method

Global claims come in various forms. It is not uncommon for the contractor to claim the recovery of all the costs incurred in the project plus profit less the amount certified and paid on the basis that, but for the matters included in the claim, the contractor would have recovered all of them. This is often referred to as a total cost method. Under the total cost method, damages are measured by comparing the claimant's actual cost of performance with its bidding estimates. The total cost method is usually used in the United States and in many ways is similar to a global claim.

This method does not attempt to make a causal connection between the claimant's damages and the defendant's actions. Instead, it assumes that all costs incurred in excess of the bid were caused by the defendant. The total cost method is based on the assumption that, if not for the defendant, then the claimant would have been able to complete the project on time and within budget. Courts have traditionally been reluctant to use the total cost method because it is based on this assumption and not on actual events. This is particularly true in complex construction disputes, where both the owner and the contractor are often responsible for a project's delays and disruptions.

In a total cost method, the contractor claims that:

- He is entitled to the difference between his original budget for labor costs, plus the amounts received in change orders for specific extra work, and its actual labor costs expended as a result of the delay.
- The exact amount of additional work performed as a result of the problems encountered is difficult, if not impossible, to determine because of the nature of the corrective work which was being performed.
- There is no precise formula by which these additional costs can be computed and segregated from those costs which he would have incurred if there had been no client caused difficulties.
- The reasonableness and accuracy of his estimate has been prepared by an experienced engineer whose qualifications will be unchallenged.

In a total cost method claim, the courts must reconcile competing interests. While courts will not prevent a claimant from recovering for a delay or disruption merely because the claimant cannot precisely quantify his actual damages, they will not give the claimant a windfall by allowing him to shift all of its overruns to the owner. Hence, the total cost method has never been totally favoured by the court and has been tolerated only when no other mode was available and when the reliability of supporting evidence was fully substantiated.

1.11 Modified Total Cost Method

The modified total cost method of calculating damages uses the total cost method as a starting point, but makes adjustments to allow for various factors (e.g., a below-cost bid, part of the claim that can be claimed separately) to arrive at a reduced figure that fairly represents the increased costs the contractor directly suffered from the particular actions of the employer.

Under this method, the claimant increases its original estimates and decreases its costs of performance to account for errors in the original bid and to account for costs that were not caused by the owner. Courts have also used a modified total cost method by substituting the court's judgment of a reasonable bid amount and deducting costs of performance that were attributable to the contractor.

The modified total cost is analogous to pricing a composite item in a schedule of work, and involves making a composite assessment in respect of a large number of causative factors. The starting point will usually be to calculate the total actual loss incurred for the entire project or relevant section as appropriate. Thereafter, adjustment and apportionment of the total loss must be made in respect of any losses which are not caused by the party in breach. This is necessary to avoid overcompensating the claimant to give effect to the underlying 'but for' principle that the claimant is entitled to be placed in the position he would have been in but for the breach of duty and the rule that the loss which is unreasonably incurred only the reasonable value was recoverable is unforeseeable and not recoverable.

The United States courts have established four criteria that a claimant must satisfy before it can use the total cost method or modified total cost method:

1. The nature of the additional costs was such that there is no other practicable means of measuring damages. Basically this criterion requires some combination of the impossibility of segregating the owner caused delay from the original cost of performance along with a showing of a lack of independent means of determining the reasonableness of contractor's expended costs.
2. The claimant's bid or estimate was realistic. In determining that a claimant's bid was reasonable, involves the bidders' qualifications, his estimates for other similar projects and the methods used and information relied on in preparing the bid.
3. The contractor costs were reasonable. The contractor has to demonstrate that it acted reasonably in incurring its additional costs. He will generally attempt to satisfy this requirement through the use of expert testimony and reliance on

industry standards. The contractor can also satisfy this criterion by demonstrating that it took measures to mitigate its additional costs.
4. The plaintiff was not responsible for any of the additional expense. This criterion is particularly difficult to satisfy in complex construction cases in which both owner and contractor usually are responsible for delays and disruptions The lack of responsibility criteria is often interpreted as a mandate to apportion fault in the total cost analysis, but it also seems to have a more fundamental purpose, and that is to assure the court of the equity of the matter that the non-breaching party receive a remedy when fault is clear.

The modified total cost method is a way of solving the problem of proving damages to a reasonable certainty, a problem to which the construction industry is particularly susceptible. It works because the courts have accepted that, when fault is certain, a non-breaching party should receive a recovery, while at the same time it protects the party in breach from runaway a damage claim. While courts continue to be somewhat hesitant about its use, it seems clear that the modified total cost method is an effective tool for achieving an equitable outcome for complicated construction disputes involving extra work claims.

References

Books

(2002) Delay and disruption protocol. Society of Construction Law, UK
Adams J, Brownsword R (1995) Key issues in contract. Butterworths, London
Adams J, Brownsword R (2007) Understanding contract law. Sweet & Maxwell, UK
Allen R, Martin S (2010) Construction law handbook. Aspen Publishers, Maryland
Ansley R, Kelleher T, Lehman A (2009) Smith, Currie & Hancock's common sense construction law: a practical guide for the construction professional. Wiley, USA
Beatson J, Burrows A, Cartwright A (2010) Anson's law of contract. Oxford University Press, Oxford
Beale H (2010) Chitty on contract. Sweet & Maxwell Ltd, UK
Beale H, Bishop W, Furmston M (2008) Contract: cases and materials. Oxford University Press, Oxford
Bockrath J, Plotnick F (2010) Contracts and the legal environment for engineers and architects. McGraw-Hill, USA
Bramble B, Callahan M (2010) Construction delay claims. Aspen Publishers, Maryland
Burrows (2005) Remedies for torts and breach of contract. Oxford University Press, Oxford
Callahan M (2010) Construction change order claims. Aspen Publishers, Maryland
Carnell N (2005) Causation and delay in construction disputes. Blackwell Publishing, Oxford
Chen-Wishart M (2010) Contract law. Oxford University Press, Oxford
Cushman R, Carter J, Gorman P, Coppi D (2007) Proving and pricing construction claims. Aspen Publishers, Maryland
Davison P, Mullen J (2008) Evaluating contract claims. Wiley-Blackwell Publishing, Oxford
Eggleston B (2008) Liquidated damages and extensions of time: in construction contracts. Wiley-Blackwell Publishing, Oxford

Furmston M, Cheshire GF, Fifoot CHS (2006) Cheshire, Fifoot and Furmston's law of contract. Oxford University Press, Oxford

Halson (2001) Contract law. Longman, London

Harris D, Campbell D, Halson R (2005) Remedies in contract and tort. Cambridge University Press, Cambridge

Hart H, Honore T (1985) Causation in the law. Oxford University Press, Oxford

Hinze J (2000) Construction contracts. McGraw-Hill, USA

Jones J (2009) Goff & Jones: the law of restitution. Sweet & Maxwell, UK

Koffman L, Macdonald E (2007) The Law of contract. Oxford University Press, Oxford

MacGregor H (2010) MacGregor on damages. Sweet & Maxwell, UK

McKendrick (2009) Contract law. Palgrave Macmillan, Basington, Hampshire

O'Sullivan J, Hilliard J (2006) The law of contract. Oxford University Press, Oxford

Peel E (2007) Treitel on the law of contract. Sweet & Maxwell, UK

Pickavance K (2010) Delay and disruption in construction contracts. Sweet & Maxwell, UK

Poole J (2010a) Textbook on contract law. Oxford University Press, Oxford

Poole J (2010b) Casebook on contract law. Oxford University Press, Oxford

Ramsey V, Furst S (2008) Keating on building contracts. Sweet & Maxwell, UK

Reese C (2010) Hudson's building & engineering contracts. Sweet & Maxwell, UK

Richards P (2009) Law of contract. Pearson Longman, Essex

Samuel G (2001) Law of obligations and legal remedies. Cavendish Publications, London

Schwartzkopf W (2004) Calculating lost labor productivity in construction claims. Aspen Publishers, Maryland

Schwartzkopf W, John J, McNamara J (2000) Calculating construction damages. Aspen Publishers, Maryland

Treitel G (2004) An outline of the law of contract. Oxford University Press, Oxford

Wallace D (1995) Hudson's building and engineering contract. Sweet & Maxwell, UK

White N (2008) Construction law for managers, architects, and engineers. Delmar Cengage Learning, USA

Wickwire J, Driscoll T, Hurlbut S, Groff M (2010) Construction scheduling: preparation, liability, and claims. Aspen Publishers, Maryland

Chapter 2
Contract Law: Principles and Doctrines

2.1 Understanding Contract Law

Doctrines in contract formation are needed to set the boundaries of proper acceptable behaviour between the parties to a contract, to save time in agreeing the principles of an agreement and to assist in dealing with common situations and reoccurring problems. They also handle the unexpected risks in the industry and deliver the required level of certainty to the whole process.

Formal doctrines in construction contract formation have become necessary in order to deal with the considerable complexity and interrelationship between the contractor, the client and the consultant as well as the specialists and the sub-contractors. This complexity arises from the split responsibility for the design and then construction, the large number of independent subcontractors, the complex price determination and procurement, the large variety of specialists and suppliers involved and the infinite number of transactions necessary to deal with the very large projects that are most common nowadays.

The doctrines are always made by a recognised and legitimate source that claimants must refer to being in writing the claim, defending it, adjudication, arbitration and court procedures. They are by order of importance:

- Statutes. Statute is an express and formal laying down of a rule or rules of conduct to be observed in the future by the persons to whom the statute is expressly, or by implication, made applicable.
- Case law, usually called common law, takes the form of court judgments. While a judgment gives reasons and may be argumentative, a statute gives no reasons and is imperative.

A. D. Haidar, *Global Claims in Construction*,
DOI: 10.1007/978-0-85729-730-3_2, © Springer-Verlag London Limited 2011

- Standard forms of contracts.[1]
- Regulations. They include building and environmental regulations, health and safety and other governmental bodies who formulate rules and regulations.
- Civil bodies such as engineers, architects, civil servants or public employees and professional institutions.
- Books, papers, journals and interviews.

Contract law, as opposed to tort, is civil rather than criminal unless injury or death by accident happens.[2] The theoretical division between contract and tort is that liability in tort arises from the breach of an obligation primarily fixed by law, whereas in contract it is fixed by the parties themselves. A *prima facie* duty of care arises if there is sufficient proximity between the alleged wrongdoer and the wronged party such that the former might reasonably expect that carelessness may cause damage to the latter. It is then necessary to consider whether there are any mitigating circumstances that reduce or limit the scope of the duty and damages.

The period of time after which a cause of action starts to run in contract is from the moment of the conduct constituting breach, whereas in tort it starts from the moment when the plaintiff sustains his damage which usually is at a much later stage.

The sources of law are forces which make the law how it is, which would therefore encompass socio-economic influences, the political process, the background and education of the judges, the behaviour of litigants and their advisers, and how cases get to court. The sources of law are the categories of material which the courts regard as authoritative in deciding cases before them. This is a definition which focuses on the courts, which in normal circumstances have the last word on what the law is, subject only to overturning by a higher court or by some higher source of law like the parliament or a similar authoritative body.

2.2   Construction Contract Law: An Introduction

The historic evolution of English and Commonwealth law, including the laws in the United States, has given central importance and authoritative power to the published or reported judgments of higher courts; so that the courts may be bound to follow principles of law whose source is no more than the decision of an earlier

---

[1] Standard form provisions are drafted to emulate common law rules. For example, financial entitlements are framed in terms of *quantum meruit* (payment for work) and damages (equivalent to breaches of contract) situations. They also allow equivalent remedies or relief to be granted within a binding contract, side stepping the undesirable consequences of the contract being avoided or terminated through common law processes.

[2] A 'tort' is defined as a legal wrong, coming from the Latin term 'torquere', which means twisted or wrong. The idea is that someone can be legally injured and tort law is used to provide restitution from another, who owes them a duty of care and can be held to be legally liable for that injury.

court of the same or higher hierarchical level. The system of judicial precedent establishes and develops principles of law from case to case and has doctrines and techniques which make this possible.

There is no single correct way to divide up the general law for the purpose of breaking it down into comprehensible components. Looking at it by its source makes little practical sense, except perhaps in relation to the distinction between law and equity. It makes more sense to use a coherent factual subject area, like perhaps construction law. However, there is no authoritative definition of what construction law covers or includes, so legal categories are a matter of choice, and hence finding the appropriate choice which is informative and appropriate for the purpose to be achieved. Construction law, together with its tables of contents, its primary sources and assuming that it covers the ground adequately, deals with law from all the primary sources discussed above.

There are real differences of focus, terminology and legal techniques between the branches of the law labelled in different ways. The main classifications that lawyers and judges have adopted in categorising law, which play an important role in how law is written and practiced cut across several other possible categories:

1. It deals about contract, tort and property. It also deals about rules or principles or remedies derived both from common law and from equity.
2. It discusses criminal law also called penal law. Criminal law involves prosecuting a person for an act that has been classified as a crime. Civil cases, on the other hand, involve individuals and organisations seeking to resolve legal disputes.
3. It encompasses substantive law. This includes the rules which lay down rights and duties or enables individuals to modify their legal position and the law of procedure which defines how rights and duties can be given effect in civil or criminal proceedings.
4. It also touches on public law, as opposed to private law, which regulates the relations between individuals and other private legal persons in relation to the powers of local authorities and other public bodies.

Many construction-related cases could get to court but do not as this is often explained by the parties' knowledge that their dispute is adequately covered by a case law rule or by a provision in a statute, so that simply litigation is not pursued. However, because the law exists through the use of language and because language is inherently imprecise, few situations offer no scope at all for arguments about what rule and doctrine applies and how automatic the outcome is in its application to a unique set of facts. There is always a scope for arguments that the apparent rule is not quite the rule it has been taken to be, or that the obvious solution from the existing authorities is actually no longer appropriate for the present situation.

2.3 Definition of a Contract

Where the law provides a framework within which the activities of the construction industry are carried out, a contract provides a sub-framework of the law for a specific undertaking. The contract arises from the agreement between two and more parties and it binds those parties in a contractual arrangement. The courts generally do not interfere between these arrangements but they will, subject to certain requirements, recognise an agreement as a contract and lend their support to enforce the agreement if called upon to do so by one of the parties. It is also possible for a contract to confer benefits on a third party enforceable by that third party.

A contract is like private law, enforceable through the state apparatus of the courts. It is sometimes said that the standard forms of contract are like private legislation even though courts in many cases tend to criticise their principles and the hardness of some of their clauses. It seems that they place more liability on the parties than the courts and the statutes do.

The essential elements of formation of a valid and enforceable contract can be summarised under the following headings:

a) There must be an offer and an acceptance which is in effect with the agreement.
b) There must be an intention to create legal relations.
c) There is a requirement of written formalities in some cases.
d) There must be consideration.
e) The parties must have capacities to contract.
f) There must be genuineness of consent by the parties to the terms of the contracts.
g) The contract must not be contrary to public policy.

The significance of a contract is that the promisee is entitled to performance of the promise or, failing that, to compensation for nonperformance. The courts will rarely enforce performance, other than payment, because it cannot or will not supervise actual performance of work. A contract by law is enforced in two stages:

1. The first stage is decision, mostly on whether there is a right to payment or compensation and how much. The law empowers the courts and arbitrators to make a binding decision on disputes properly submitted to them.
2. The second stage is that a court can enforce a judgment or arbitration award through various processes such as the appointment of a receiver and seizure of goods.

Although the definition of a contract refers to promises, the primary obligations will, in some transactions, be fulfilled at the time of contract. The obligations in such a case are said to be executed, whereas if an obligation is to be performed in the future, it is said to be executory. Where obligations are executory, the usual position is that both parties are mutually bound to perform. Traditionally, courts have distinguished between bilateral and unilateral contracts by determining whether one or both parties provided consideration and at what point they provided the consideration.

Bilateral contracts were said to bind both parties the minute the parties exchange promises, as each promise is deemed sufficient consideration in itself. Unilateral contracts, also called 'if' contracts, are said to bind only the promisor and do not bind the promisee unless the promisee accepts by performing the obligations specified in the promisor offer. Until the promisee performs, he or she has provided no consideration under the law. A typical example of a unilateral contract in construction is an offer for completing a project at a specified time earlier than the completion date. The question whether the promisor under an 'if' contract is always bound to keep open the promise once the promisee embarks upon relevant acts is a moot point. The promisee may be entitled to some remedy such as a *quantum meruit* if the offer is withdrawn in such circumstances.

In *Errington v Errington & Woods*,[3] the court held, *obiter dicta*, that within a unilateral contract, there is an implied promise not to revoke the contract once performance has commenced and protects the interest of the party who is acting on the promise of the offeror. Lord Denning stated the following: "It could not be revoked by him once the couple entered on performance of the act, but it would cease to bind him if they left it incomplete and unperformed."

In *British Steel Corp. v Cleveland Bridge*,[4] the court held that British Steel were entitled to £200,000 for work done, on a quasi-contractual basis, but as there was no concluded contract, Cleveland Bridge counterclaim for £800,000, for late delivery and delivery out of sequence, failed. In his judgment, J. Robert Goff defined an 'if' contract by stating:

> There may be what is sometimes called an 'if' contract, i.e. a contract under which A requests B to carry out a certain performance and promises that, if he does so, he will receive a certain performance in return, usually remuneration for his performance. The latter transaction is really no more than a standing offer which, if acted on before it lapses or is lawfully withdrawn, will result in a binding contract. The question whether any contract has come into existence must depend on a true construction of the relevant communications which have passed between the parties and the effect (if any) of their actions pursuant to those communications.

In summary, there are three main requirements for a legally valid contract which the law will enforce:

- Intention to create legal relations.
- Agreement and essential terms.
- Consideration.

---

[3] (1952) 1 KB 290. This case concerned a contract by a father to allow his son to buy the father's house on payment of the instalments of the father's building society loan.

[4] British Steel Corporation v Cleveland Bridge & Engineering Co Ltd (1984) 1 All ER 504. Cleveland Bridge was the contractor for a steel framed building in Damman. They contacted British Steel to supply some special cast steel nodes and issued a letter of intent. As the work was urgent, British Steel started immediately and in fact completed and delivered all 137 nodes with arguments over the price and other terms still unresolved.

2.4 Formation of Contracts

The requirements and related rules are important in determining whether a contract exists. In some situations one or another party may seek to prove that either there is or there is not a contract, in order to secure a benefit or avoid a liability. The courts appear to adopt two different approaches, either an open-minded approach as to whether a contract exists, or starting from a presumption that a contract exists. Under English law, which most of the current contract law is based upon, an agreement supported by consideration is not enough to create a legally binding contract; the parties must also have an intention to create legal relations. Often, the intention to create legal relations is expressly stated by the contracting parties. In other situations, the law will readily imply the intention because of the nature of the contractual dealings between the parties.

In construction agreements, there is a rebuttable presumption that parties intend to create legal relations and conclude a contract. In determining whether parties have created legal relations, courts will look at the intentions of the parties. If in the course of the business transactions, the parties clearly and expressly make an agreement stating that it ought not to be binding in law, then the court will uphold those wishes. However, if a court is of the view that there is any ambiguity of intention, or that such an intention is unilateral, the contract will be voided. The burden of rebutting the presumption of legal relations in construction agreements lies on the party seeking to deny the contract.

The presence of consideration is often indicative of the intention to create legal relations, though there are situations where the presumption of the intention can be rebutted, thus determining that there is no contract and no legal liability.

The courts will examine sometimes in construing if a contract exists based on the maxim of 'validate if possible' which is expressed by '*verba ita sunt intelligenda ut res magis valeat quam pereat*' as explained by Lord Wright in *Hillas & Co Ltd v Arcos Ltd*[5]:

> But it is clear that the parties both intended to make a contract and thought they had done so. Businessmen often record the most important agreements in crude and summary fashion; modes of expression sufficient and clear to them in the course of their business may appear to those unfamiliar with the business far from complete or precise. It is accordingly the duty of the court to construe such documents clearly and broadly without being too astute or subtle in finding defects; but, on the contrary the court seeks to apply the old maxim of English law, verba ita sunt intelligenda ut res magis valeat quam pereat. That maximum, however, does not mean that the court is to make a contract for the parties, or to go outside the words they have used, except in so far as there are appropriate implications or law, as for instance, the implication of what is just and reasonable to be ascertained by the court as a matter of machinery when the contractual intention is clear but the contract is silent on some details.

---

[5] (1932) 147 LT 503. A contract relating to timber containing an option clause did not specify what kind or sizes or quantities were to be supplied, nor did it define the dates and ports of shipment and discharge.

In *Courtney & Fairbairn v Tolaini Brothers (Hotels) Ltd*,[6] where it was held that the plaintiff could not recover any loss of profit as there was no concluded contract. Lord Denning stated:

> If the law does not recognise a contract to enter into a contract (when there is a fundamental term yet to be agreed) it seems to me it cannot recognise a contract to negotiate. The reason is because it is too uncertain to have any binding force. No court could estimate the damages because no one can tell whether the negotiations would be successful or would fall through; or if successful, what the result would be. It seems to me that a contract to negotiate, like a contract to enter into a contract, is not a contract known to the law.

In *The Aramis*,[7] LJ Bingham considered the authorities at some length to see how the implication of contracts in this field had grown and developed. He cited with approval from the judgment of LJ May in *The Elli*[8] which said:

> As the question whether or not any such contract is to be implied is one of fact, its answer must depend upon the circumstances of each particular case—and the different sets of facts which arise for consideration in these cases are legion. However, I also agree that no such contract should be implied on the facts of any given case unless it is necessary to do so: necessary, that is to say, in order to give business reality to a transaction and to create enforceable obligations between parties who are dealing with one another in circumstances in which one would expect that business reality and those enforceable obligations to exist.

LJ Bingham then continued himself to say:

> Whether a contract is to be implied is a question of fact and that a contract will only be implied where it is necessary to do so. It would, in my view, be contrary to principle to countenance the implication of a contract from conduct if the conduct relied upon is no more consistent with an intention to contract than with an intention not to contract. It must, surely, be necessary to identify conduct referable to the contract contended for or, at the very least, conduct inconsistent with there being no contract made between the parties. Put another way, I think it must be fatal to the implication of a contract if the parties would or might have acted exactly as they did in the absence of a contract.

In some situations, a contract may be implied from acts and circumstances. In *Blackpool Aero Club v Blackpool BC*,[9] the Court of Appeal held, *obiter dicta*,

---

[6] (1975) 1 All ER 716. Tolaini Brothers owned a site and proposed to have built a hotel and other projects. Courtney & Fairbairn, who were builders, proposed that if they could introduce finance, they would be employed as the builder with 'fair and reasonable contract sums' to be negotiated in respect of each project as it arose 'based upon agreed estimates of the net cost of the work and general overheads with a margin of profit of 5%.' Finance was successfully introduced by Courtney & Fairbairn but the negotiations and the agreement of estimates failed. Another builder was brought in.

[7] The Aramis (Cargo Owners) v Aramis (Owners) (1989) 1 Lloyd's Rep 213.

[8] Ilyssia Compania Naviera SA v Bamaodah, The Elli 2 (1985) 1 Ll R 107.

[9] Blackpool and Fylde Aero Club v Blackpool Borough Council (1990) 3 All ER 25. In this case a group of six tenderers were invited to tender for the concession to operate pleasure flights from the local airport. The invitation to tender specified a deadline and indicated that no late tenders would be accepted. The plaintiffs submitted a tender prior to the deadline but the defendant did not empty their post box as a result of which it was not considered. The court held that a contract should be implied which required the defendant to open and consider the plaintiff's tender in conjunction with all other tenders.

that the council was contractually obliged to give proper consideration to complying tenders. This contract consisted of an offer contained in the tender advertisement which was accepted by each tenderer who put in a complying tender. The
promise to give proper consideration to complying tenders was not explicit but the
court was prepared to find an implied promise in those terms. This finding was
made despite the fact that in other respects the council had explicitly excluded the
possibility of legal obligation, for example, it had reserved the right not to accept
any tender. The court drew support for its implication of a term from the fact that
this tender was conducted according to a 'clear, orderly and familiar procedure'.
The way the tender was conducted gave rise to the expectation that it would be
conducted properly. This argument tends to reinforce the view that undertakings
given by governmental bodies in tender advertisements should give rise to
enforceable contractual obligations.

Thus in *Blackpool Aero Club v Blackpool BC*, the invitation of tenders from
selected people, subject to specific requirements, and their submission of tenders in
accordance with those requirements gave rise to an implied contract. In the
absence of clearly defined rights of tenderers, clients could, at least in theory, have
behaved largely as they liked. LJ Bingham described the potential unfairness of
the situation, which he saw as "heavily weighted in favour of the invitor". He said
the following:

> I readily accept that contracts are not to be lightly implied. Having examined what the
> parties said and did, the court must be able to conclude with confidence both that the
> parties intended to create legal relations and that the agreement was to the effect contended
> for. The council's invitation to tender was, to this limited extent, an offer and the club's
> submission of a timely and conforming tender an acceptance.

In *Percy Trentham v Archital Luxfer*,[10] engineering work was completed
without a written contract. The Court of Appeal held that, despite the parties
failure to execute a formal contract; the commercial nature of the transaction, and
the fact that the work had been fully carried out both showed that there was indeed
a contract and that the subcontract came into existence "not simply by an
exchange of correspondence but partly by reason of written exchanges, partly by
oral discussions and partly by performance of the transactions."

In *Percy Trentham v Archital Luxfer*, LJ Staughton identified the mechanism
for contract formation by stating the following:

> Before I turn to the facts it is important to consider briefly the approach to be adopted to
> the issue of contract formation in this case. It seems to me that four matters are of
> importance. The first is the fact that English law generally adopts an objective theory of

---

[10] G. Percy Trentham Ltd v Archital Luxfer Ltd (1993) 1 Lloyd's Rep 25. Trentham were main
contractors for the construction of industrial units. Archital were the aluminum walling
subcontractors. Archital submitted a number of quotations and there was then a round of offer and
counter-offer, but a formal subcontract was never signed. The subcontract work was fully
performed, but performed late with the result that the employer sought and obtained damages
from Trentham, who then sought an indemnity from Archital. Archital denied there was a
contract.

contract formation. That means in practice our law generally ignores the subjective expectations and the unexpressed mental reservations of the parties. Instead the governing criterion is the reasonable expectations of honest men. And in the present case that means that the yardstick is the reasonable expectations of sensible businessmen. Secondly, it is true that the coincidence of offer and acceptance will in the vast majority of cases represent the mechanism of contract formation. It is so in the case of a contract alleged to have been made by an exchange of correspondence. But is not necessarily so in the case of a contract alleged to have come into existence during and as a result of performance. The third matter is the impact of the fact that the transaction is executed rather than executory. It is a consideration of the first importance on a number of levels. The fact that the transaction was performed on both sides will often make it unrealistic to argue that there was no intention to enter into legal relations. It will often make it difficult to submit that the contract is void for vagueness or uncertainty. Specifically, the fact that the transaction is executed makes it easier to imply a term resolving any uncertainty, or, alternatively, it may make it possible to treat a matter not finalised in negotiations as unessential. In this case fully executed transactions are under consideration. Clearly, similar considerations may sometimes be relevant in partly executed transactions. Fourthly, if a contract only comes into existence during and as a result of performance of the transaction it will frequently be possible to hold that the contract impliedly and retrospectively covers pre-contractual performance.

The use of partnering arrangements is now widespread in the construction industry. It is generally thought that partnering agreements do not in themselves constitute contract obligations. The partnering objectives are more in the nature of management ideals and therefore sit alongside the formal contract undertakings. The principles to be applied in a partnering arrangement are:

1. The intention of the parties to agree in the relevant period of negotiation.
2. At the time of the contract, the agreement include with sufficient certainty the terms required in order that a contract should come into existence.
3. The terms include all the terms which were in fact essential to be agreed if the contract was to be legally enforceable and commercially workable.
4. The existence of sufficient acceptance by the offeree of the offer, to comply with any stipulation in the offer itself as to the manner of acceptance.

2.5 Elements of an Agreement

The second requirement for a legally binding contract is agreement of the essential terms. This has two aspects mainly what is an agreement and what are the essential terms of an agreement. As to an agreement the old legal concept of *consensus ad idem*, a meeting of minds, has been replaced by the objective approach that an agreement is not a mental state but an act, and as an act, is a matter of inference from conduct where the parties are to be judged, not by what is in their minds, but by what they have said or written or done.

Agreement may be evidenced by a single instrument, which may, but need not necessarily, be signed or executed as a deed by both parties. Where the agreement is set out in a single instrument, this agreement supersedes previous negotiations.

It is equally possible for a binding agreement to be found in exchanges between the parties. It is then common to analyse the process of agreement in terms of offer and acceptance, on the basis that agreement is the unqualified acceptance of an offer. In the construction industry established mechanism, a clear and simple offer and acceptance process is found by the system of inviting tenders.

Acceptance may be in the form of express words, either spoken or written; but equally it may be implied from conduct as can be seen from *Brogden v Metropolitan Railway Co.*[11] For an agreement to be inferred from conduct, the conduct must be referable to some agreement or draft agreement, which was in existence before the contract started. Whether a binding contract has been established by the conduct of the parties is a question of fact.

As Lord Cozens-Hardy MR, in *Perry v Suffields*,[12] said in confirming the basic elements of an agreement and the effect of offer and acceptance on the validity of the contract:

> Though when a contract is contained in letters, the whole correspondence should be looked at, yet if once a definite offer has been made, and it has been expected without qualification, and it appears that the letters of offer and acceptance contained all the terms agreed on between the parties, the complete contract thus arrived at cannot be affected by subsequent negotiations. When once it is shown that there is a complete contract, further negotiations between the parties cannot, without the consent of both, get rid of the contract already arrived at.

Hillas v Arcos (1932) (*supra*) is a landmark House of Lords case on contract law where the court first began to move away from a strict, literal interpretation of the terms of a contract, and instead interpreted it with a view to preserve the bargain. The court ruled that judges may infer terms into a contract based on the past dealings of the parties rather than void the agreement. Lord Wright noted that courts must interpret contracts "fairly and broadly" following the maxim that "words are to be so understood that the subject-matter may be preserved rather than destroyed." Lord Wright qualified this statement by saying that courts can never create a contract where there is none. He further said:

---

[11] (1877) 2 App Cas 666. The claimants were the suppliers of coal to the defendant railway company. They had been dealing for some years on an informal basis with no written contract. The parties agreed that it would be wise to have a formal contract written. The defendant drew up a draft contract and sent it to the claimant. The claimant made some minor amendments and filled in some blanks and sent it back to the defendant. The defendant then simply filed the document and never communicated their acceptance to the contract. Throughout this period the claimants continued to supply the coal. Subsequently a dispute arose and it was questioned whether in fact the written agreement was valid.

[12] (1916) 2 Ch 187. In *Perry v Suffields*, the vendor was granted specific performance of a contract contained in two letters of February 23 and March 3, 1915. The defendant's solicitors sent a draft agreement in a letter in which they stated, in part: 'We do not know whether it incorporates quite all the terms agreed, as Mr. Perry has not seen it and we have not had very full instructions from him.' The draft contract contained clauses at variance with that agreed upon and when it was contended that this amounted to a reopening of the arrangement between the parties.

The object of the court is to do justice between the parties and the court would do its best, if satisfied that there was an ascertainable and determinate intention to contract, to give effect to such intention looking at substance and not mere form. It would not be deterred by mere difficulties of interpretation. Difficulty is not synonymous with ambiguity so long as any definite meaning can be extracted. The test of intention is to be found in the words used. If these words considered however broadly and untechnically and with due regard to all the just implications failed to evince any definite meaning on which the court can safely act, the court has no choice but to say that there is no contract. Such a position is not often found.

In *Foley v Classique Coaches Ltd*,[13] the court held that a mere agreement to agree is regarded as uncertain, unless there is an adequate machinery to resolve any lack of agreement as clearly stated by J. Maugham: "It is indisputable that unless all the material terms of the contract are agreed, there is no binding obligation. An agreement to agree in the future is not a contract; nor is there a contract if a material term is neither settled nor implied by law and the document contains no machinery for ascertaining it."

In *Nicolene Ltd v Simmonds*,[14] steel bars were bought on terms which were certain except for a clause that the sale was subject to '*the usual conditions of acceptance*'. There being no such usual conditions, it was held that the phrase was meaningless, but that this did not vitiate the whole contract; the words were severable and could be ignored. Viscount Maugham said: "In order to constitute a valid contract the parties must so express themselves that their meaning can be determined with a reasonable degree of certainty. It is plain that unless this can be done it would be impossible to hold that the contracting parties had the same intentions; in other words consensus ad idem would be a matter of mere conjecture."

Where a contract is concluded after work has started, it refers back to the commencement of the work. In *Trollope & Colls Ltd v Atomic Power Constructions Ltd*,[15] it was found that, irrespective of when the contract between the parties was signed, it did relate back so as to apply to what had been performed. In that case the court found that at the date a formal contract came into existence, there had been an intention to make a contract, there was agreement on all essential terms and a sufficiently clear acceptance of the offer, and that, therefore, a term should be implied to give business efficacy to the agreement to the effect that the terms applied retrospectively. J. Megaw stated that: "There was no principle of English law which provided that a contract cannot in any circumstances have

---

[13] (1934) 2 KB 1. The defendants operated a fleet of motor coaches, and purchased a piece of land from the plaintiffs, who operated a service station on adjacent premises. The defendants entered into a supplemental agreement to purchase all of their required fuel from the plaintiff's at a price that would be periodically agreed upon by the parties. After three years of the defendants obtaining all of their fuel from the plaintiffs, they attempted to repudiate the supplemental agreement. The plaintiffs sought a declaration that the agreement was binding and an injunction to prevent the defendants from purchasing their petrol elsewhere.

[14] (1953) 1 Q.B. 543.

[15] (1962) 3 All ER 1035.

retrospective effect. Often the contract expressly so provides. I can see no reason why, if the parties so intend and agreed, such a stipulation should be denied legal effect."

In *Port Sudan Cotton Co v Govindaswamy Chettiar & Sons*,[16] the court held, *obiter dicta*, that the parties are to be regarded as masters of their contractual fate; it is their intentions which matter and to which the court must strive to give effect. In this endeavour, the observation of Lord Denning provides much help as he stated:

> In considering this question, I do not much like the analysis in the text-books of enquiring whether there was an offer and acceptance, or a counter-offer, and so forth. I prefer to examine the whole of the documents in the case and decide from them whether the parties did reach an agreement upon all material terms in such circumstances that the proper inference is that they agreed to be bound by those terms from that time onwards.

In *Pagnan SpA v Feed Products Ltd*,[17] the court held that a contract had been concluded because the agreement had been reached on all matters which the parties themselves regarded as essential. The outstanding matters were regarded by the parties as relatively minor details which could be sorted out once a bargain had been struck. J. Bingham said:

> The general principles to be applied in deciding the issue in this case are to, I think, open to much doubt. The court's task is to review what the parties said and did and from that material to infer whether the parties objective intentions as expressed to each other were to enter into a mutually binding contract. The court is not of course concerned with what the parties may subjectively have intended.

This view was endorsed by the Court of Appeal where LJ Lloyd commented: "There was no legal obstacle which stands in the way of the parties agreeing to be bound now while deferring important matters to be agreed later."

In *Pagnan SpA v Feed Products Ltd*, LJ Lloyd sets out a simple test where a failure to agree on a term may not invalidate a contract unless thereby the contract becomes unworkable; and whether a term is so necessary as to be essential, whereby a failure to agree on it precludes an agreement from being binding in law or makes a contract unworkable, is a question for the parties to decide. In that respect he stated the following:

> (1) In order to determine whether a contract has been concluded in the course of correspondence, one must first look to the correspondence as a whole. (2) Even if the parties have reached agreement on all the terms of the proposed contract, nevertheless they may intend that the contract shall not become binding until some further condition has been fulfilled. This is the ordinary 'subject to contract' case. (3) Alternatively, they may intend that the contract shall not become binding until some further term or terms have been agreed. (4) Conversely, the parties may intend to be bound forthwith even though there are

---

[16] (1977) 2 Lloyds Rep 5.

[17] (1987) 2 Lloyd's Rep 601. Intending sellers of a quantity of corn gluten pellets subsequently contended that a contract negotiated through an intermediary had not been concluded on the grounds that a number of matters had not been agreed and negotiations were continuing.

further terms still to be agreed or some further formality to be fulfilled. (5) If the parties fail to reach agreement on such further terms, the existing contract is not invalidated unless the failure to reach agreement on such further terms renders the contract as a whole unworkable or void for uncertainty. (6) It is sometimes said that the parties must agree on the essential terms and that it is only matters of detail that can be left over. This may be misleading, since the word 'essential' in that context is ambiguous. If by 'essential' one means a term without which the contract cannot be enforced then the statement is true; the law cannot enforce an incomplete contract. If by 'essential' one means a term which the parties have agreed to be essential for the formation of a binding contract, then the statement is tautologous. If by 'essential' one means only a term which the Court regards as important as opposed to a term which the Court regards as less important or a matter of detail, the statement is untrue. It is for the parties to decide whether they wish to be bound and, if so, by what terms, whether important or unimportant. It is the parties who are, in the memorable phrase coined by the Judge, 'the masters of their contractual fate'. Of course, the more important the term is the less likely it is that the parties will have left it for future decision. But there is no legal obstacle which stands in the way of the parties agreeing to be bound now while deferring important matters to be agreed later. It happens everyday when parties enter into so-called 'heads of agreement'.

The law requires for an agreement to be a binding contract, not only that all essential terms should be agreed, but also that the terms agreed should be sufficiently certain. This is distinct from the question whether the terms eventually agreed can be identified, such as arose in *Percy Trentham v Archital Luxfer* (1993) (*supra*). The courts will not make an agreement for the parties. On the other hand, the courts will try to give meaning to a contract if they find the parties intended to be bound. Repugnant or surplus words may be rejected or modified, or clearly intended words supplied, in order to save a document.

2.6 Offer and Acceptance

Although it is not essential to do so, it is conventional to analyse an agreement, which has not been reduced to a single document, in terms of offer and acceptance. Two basic principles are applied where an agreement is concluded by the unequivocal and unconditional acceptance of a specific offer and the fact that a counter-offer kills off an offer.

These principles are used not only to decide when and whether a contract has come into existence, but also what terms have survived and emerged as part of the contract. The general rules to offer and acceptance are:

1. Standing offer can be revoked at any time only by a party making the offer and only before acceptance by other party.
2. The effect of an offer is that it confers on the offeree the power to accept it and thereby to create a binding contract.
3. An acceptance outside the time for valid acceptance will be too late. If no time is specified, the acceptance must be made within a reasonable time.
4. The ordinary rule at common law is that an offer is accepted when the offeror receives notice of the acceptance.

The basic principles are that introducing a new term, or deleting a term, is a counter-offer. The effect of the counter-offer in law is to kill the original offer so that it cannot then be accepted unless revived. But a mere request for further information is not a counter-offer. This principle enables the outcome of a sequence of exchanges to be analysed, both to determine when and whether a contract has come into existence and what the outcome is.

In *Hyde v Wrench*,[18] an offer of the sale of property was answered with a counter-offer of a lower amount. This counter-offer destroyed the first offer so it could not later be accepted. This case is a simple illustration of the proposition that a counter-offer is usually taken as rejecting the earlier offer. It was held that there was no contract. In a situation where a purported acceptance introduces new terms, no contract is formed, the initial offer has been rejected and a counter-offer has been made. According to Lord Langdale MR:

> I think there exists no valid binding contract between the parties for the purchase of the property. The defendant offered to sell it for £1,000, and if that had been at once unconditionally accepted, there would undoubtedly have been a perfect binding contract; instead of that, the Plaintiff made an offer of his own, to purchase the property for £950, and he thereby rejected the offer previously made by the Defendant. I think that it was not afterwards competent for him to revive the proposal of the Defendant, by tendering an acceptance of it; and that, therefore, there exists no obligation of any sort between the parties.

An offer, to be capable of acceptance, must indicate an intention to be bound. A distinction is made between an offer and a mere an invitation to treat. In *Carlill v Carbolic Smoke Ball Co*,[19] it was also said that the contract is made with the world; that is, with everybody, and that you cannot contract with everybody. According to LJ Bowen:

> One cannot doubt that, as an ordinary rule of law, an acceptance of an offer made ought to be notified to the person who makes the offer, in order that the two minds may come together. Unless this is done the two minds may be apart, and there is not that consensus which is necessary, according to law, to make a contract. But there this clear gloss to be made upon that doctrine, that as notification of acceptance is required for the benefit of the person who makes the offer, the person who makes the offer may dispense with notice to himself if he thinks it desirable to do so, and I suppose there can be no doubt that where a person in an offer made by him to another person, expressly or impliedly intimates a particular mode of acceptance as sufficient to make the bargain binding, it's only necessary for the other person to whom such offer is made to follow the indicated method of acceptance; and if the person making the offer, expressly or impliedly intimates in his offer

---

[18] (1840) 3 Beav 334. A offered to sell B a property for £1,000. B, in his response, offered £950, which A refused. Then B agreed to give A £1,000, but A refused to sell. B sued for specific performance of the alleged contract. The court held that B's offer to buy at £950 in response to the offer was a refusal followed by a counter-offer, and that no contract was formed.

[19] (1892) 2 QB 484. D advertised the smoke balls with an offer to pay £100 to anyone who succumbed to influenza after using a smoke ball. The advert stated that £1,000 had been deposited with bankers. P bought and used a smoke ball but caught influenza. It was argued by D that the advert was not meant to be taken seriously, but it was held that P was entitled to the £100.

that it will be sufficient to act on the proposal without communicating acceptance of it to himself, performance of the condition is a sufficient acceptance without notification.

The case is also used as an example of an 'if' contract. The offer is accepted by the other party doing what was required which in this case using the carbolic smoke ball.

In *Harvela Investments Ltd v Royal Trust Co of Canada*,[20] it was held that vendors, by inviting sealed tenders for shares, assumed a binding legal obligation to enter into a bilateral contract with the highest fixed price tenderer; therefore, where sealed tenders are invited, this may constitute an offer. The analysis which preserves the theory that the bidder makes an offer and, at the same time, gives force to a promise which is about the conduct of the competition and a promise to sell to the highest bidder in fixed bid cases, is that the promise is an offer of a unilateral type which invites a certain response, usually to put in a complying bid. Once the bidder has responded appropriately, then there is a contractual obligation on the part of the seller to sell to the highest bidder. The unilateral contract is, in effect, a contract to make a contract.

The significance of the *Harvela Investments Ltd v Royal Trust Co of Canada* decision is large. It has particular implications for tendering, the most common way in which governments and large companies make important purchases and make contractual arrangements for their works. On the traditional analysis the process or conduct of a tender is not governed by any legal obligations. This is because, again, there is no contract until a particular tender has been accepted, each tender being an offer. However, it is quite possible to argue that there is a contract before the main contract. This contract before a contract governs the way in which the tender is conducted and according to Lord Diplock: "The mere use by the vendors of the words offer was not sufficient. The task of the court is to construe the invitation and to ascertain whether the provisions of the invitation, read as a whole, create a fixed bidding sale."

In *Blackpool & Fylde Aero Club v Blackpool* (1990) (*supra*), the Court of Appeal held that the council was contractually obliged to give proper consideration to complying tenders. This contract consisted of an offer contained in the tender advertisement which was accepted by each tenderer who put in a complying tender. The promise to give proper consideration to complying tenders was not explicit but the court was prepared to find an implied promise in those terms. LJ Bingham upheld:

> The council invitation was, to this limited extent, an offer and the club's submission of a timely and conforming tender an acceptance. The invitee is in my judgment protected at least to this extent: if he submits a conforming tender before the deadline he is entitled, not as a matter of mere expectation but of contractual right to be sure that his tender will after the deadline be opened and considered in conjunction with all other conforming tenders or at least that his tender will be considered if others are.

---

[20] Harvela Investments Ltd v Royal Trust Co. of Canada (C.I.) Ltd (1986) AC 207.

The Court of Appeal did not propose to remove all these risks, nor even to address them. But it did offer for the first time a degree of protection for the tenderer in cases where a conforming bid was wrongly rejected. According to LJ Bingham: "The invitee is in my judgment protected at least to this extent: if he submits a conforming tender before the deadline he is entitled, not as a matter of mere expectation *but of contractual right* to be sure that his tender will after the deadline be opened and considered in conjunction with all other conforming tenders or at least that his tender will be considered if others are."

In *Aries Power Plant v ECE Systems Ltd*,[21] the court held, *obiter dicta*, that the concept of the counter-offer killing the offer may be an overstatement, where they are not wholly incompatible. HHJ Humphrey Lloyd presented a modified approach to dealing with a sequence of correspondence. He stated: "As the parties are agreed on the documents which constitute the contract, it is in my judgment necessary to determine the issues on the basis of the documents. In any event I do not accept the argument that I should determine which of these documents constituted an offer or counter-offer and then ignore the provisions of the offer which was superseded by the counter-offer. This was not a case in which there was a 'battle of forms' or a series of communications, each of which was intended to supplant a previous communication."

2.7 Consideration

The third requirement for a binding contract is that a promise is only binding if it is given for good consideration, or if it is executed as a deed. Valuable consideration has been defined as some right, interest, profit, or benefit accruing to the one party, or some forbearance, detriment, loss, or responsibility given, suffered, or undertaken by the other at his request. It is not necessary that the promisor should benefit by the consideration. It is sufficient if the promisee does some act from which a third person benefits, and which he would not have done but for the promise.[22]

Therefore, consideration for a promise may consist in either some benefit conferred on the promisor, or detriment suffered by the promisee, or both. On the other hand, that benefit or detriment can only amount to consideration sufficient to support a binding promise where it is causally linked to that promise. The purpose of the requirement of consideration is to put some legal limits on the enforceability of agreements even when they are intended to be legally binding and are not vitiated by some factor such as mistake, misrepresentation, duress or illegality. The traditional definition of consideration concentrates on the requirement that something of value must be given and accordingly states that consideration is either some detriment to the promise in that he may give a value, or some benefit

---

[21] (1995) 45 Con LR 111.
[22] Currie v Misa (1875) LR 10 Exch 153.

to the promisor in that he may receive a certain value. Law is concerned with consideration for a promise not the consideration for a contract.

The classic definition of consideration was provided in a statement by Sir Frederick Pollock, adopted by Lord Dunedin in *Dunlop v Selfridge*[23]: "An act or forbearance of one party, or the promise thereof is the price for which the promise of the other is bought, and the promise thus given is enforceable."

Consideration is usually either money or a promise to pay money, but it can also be some other benefit to the promisor or detriment to the promisee. In *Shanklin Pier Ltd v Detel Products Ltd*,[24] where it was held that paint specification for purchase by a contractor to use on the pier was sufficient consideration for the warranty that the paint was fit for the purpose. The court held, *obiter dicta*, that on the facts the representation was a warranty; the consideration for the warranty was that the plaintiffs should cause the contractors to enter into a contract with the manufacturers for the supply of the paint for re-painting the pier; and, therefore, the warranty was enforceable and the paint manufacturers were liable in damages for its breach. J. Mcnair stated: "If, as is elementary, the consideration for the warranty in the usual case is the entering into of the main contract in relation to which the warranty is given, I see no reason why there may not be an enforceable warranty between A and B supported by the consideration that B should cause C to enter into a contract with A or that B should do some other act for the benefit of A."

In *Williams v Roffey Bros & Nicholls (Contractors) Ltd*,[25] the Court of Appeal held that an agreement between a main contractor and a subcontractor to make an additional payment to the subcontractor to ensure that he continued with the work and completed on time was binding. In the absence of fraud or economic duress and in the particular situation where the subcontractor would otherwise have been unable to continue work, there was a benefit which amounted to consideration. The general rule was stated, nevertheless, to be a good law. LJ Glidewell said:

> (1) if A has entered into a contract with B to do work for, or to supply goods or services to, B in return for payment by B and (2) at some stage before A has completely performed his obligations under the contract B has reason to doubt whether A will, or will be able to, complete his side of the bargain and (3) B thereupon promises A an additional payment in return for A's promise to perform his contractual obligations on time and (4) as a result of giving his promise B obtains in practice a benefit, or obviates a disbenefit, and (5) B's promise is not given as a result of economic duress or fraud on the part of A, then (6) the

---

[23] Dunlop Pneumatic Tyre Co Ltd v Selfridge & Co Ltd (1915) AC 847.

[24] (1951) 2 All ER 471. Paint manufacturers represented to the owners of a pier that a paint which they manufactured was suitable for re-painting the pier. In reliance on this representation the pier owners specified that contractors under a contract with them to re-paint the pier should use the paint. The paint proved to be a failure and the pier owners suffered loss in consequence.

[25] (1991) 1 QB 1. A sub-contract has been made between the plaintiff and the contractors to do carpentry works in a block of 27 flats. Unfortunately, some problems had raised between the two parties hence, the sub-contractor, Williams, ceased work. The latter then sued the defendant for breaching of contract. The main problems in this case were whether or not there was consideration.

benefit to B is capable of being consideration for B's promise, so that the promise will be legally binding.

The current judicial tendency is to minimise the impact of the doctrine and to find that consideration exists in quite minor benefits or detriments. For example, in *Barry v Heathcote Ball & Co (Commercial Auctions) Ltd*,[26] the Court of Appeal had to explain what consideration was given in return for the auctioneer's promise that the sale would be without reserve. Per Sir Murray Stuart-Smith: "As to consideration, in my judgment there is consideration both in the form of detriment to the bidder, since his bid can be accepted unless and until it is withdrawn, and benefit to the auctioneer, as the bidding is driven up. Moreover, attendance at the sale is likely to be increased if it is known that there is no reserve."

2.8 Implied Terms

Terms may be classified according to a number of characteristics. The first is that terms may be express or implied. Express terms are those terms to be derived from the written or spoken words, actions and/or conduct forming the agreement. Terms are implied by law to supplement the express terms. Standard terms will be implied to avoid the need for the parties to express each time that which goes without saying. Specific terms may be implied to fill in gaps in the parties' agreement.

The implication of terms in specific transactions is governed by different principles. A term will be implied where necessary to give business efficacy to the agreement and to give effect to the presumed intentions of the parties. A term will readily be implied into a contract; that the parties shall co-operate to ensure the performance of their bargain. However, the legal concept of co-operation is restricted to an obligation firstly to perform those duties which must be performed to enable the other party to perform his own obligations under the contract, and secondly not to prevent the other party from performance.

The conventional lawyer analysis of contracts is to identify terms of the contract. The meaning of this is explained that "if a statement is a term of the contract, it creates a legal obligation for whose breach an appropriate action lies at common law."[27] In the event of a breach of a term of a contract the basic right which a party derives from the contract is to bring an action, that is, to sue upon the breach and recover money damages in respect of loss or damage suffered as a result of the breach.

Various tests have been proposed to identify whether a pre-contract statement is to be regarded as a mere representation or imposing a contractual liability. Chitty[28]

---

[26] (2001) 1 All ER 865.

[27] Cheshire, Fifoot and Furmston—Law of Contract.

[28] Chitty on Contract.

concludes the only true test is "whether there is evidence of an intention by one party or the other that there should be contractual liability in respect of the accuracy of the statement". Representations may be incorporated as terms in the contract, but pre-contract statements intended as 'mere representations', i.e., not incorporated as terms of the contract, did not give rise to liability to damages at common law in the absence of fraud or negligence.[29]

The object of construction of the terms of a written agreement is to discover, in the light of the matrix of facts known to the parties at the time of contract, the intention of the parties to the agreement. As per Lord Wilberforce in *Prenn v Simmonds*[30]:

> As in all cases of construction of written contracts, this exercise (of interpreting the contract) is to be carried out, not in vacuo but against the background of the surrounding circumstances or matrix of facts known to the parties at or before the agreement insofar as those throw light on what was the commercial or business object of the transaction objectively ascertained. In order for the agreement to be understood, it must be placed in its context. The time has long passed when agreements, even those under seal, were isolated from the matrix of facts in which they were set and interpreted purely on internal linguistic considerations.

The basis on which the courts can imply terms is controversial. Traditionally, the justification has been that they were giving effect to the presumed intention of the parties. In these cases the idea is that it is necessary to presume the term was intended otherwise the contract would not function properly. It is possible to divide terms incorporated by the courts into two categories; terms implied in fact and terms implied in law. In the latter case, the term will be implied in every contract of that kind, whereas in the former case the term will only be implied where the facts of the case give rise to a need, i.e., where there is evidence that it was an unexpressed intention of the parties. The importance of the distinction is that the test seems to be lower for a term to be implied in law, where it is only reasonably necessary, whereas for terms implied in fact it must be necessary.

However, the distinction between terms implied in law and in fact is not always easy to draw. The case law below reveals two main tensions: the first is whether there should be a distinction between terms in law and in fact and the second is whether the overall test for implying a term should be based on reasonableness or on necessity.

The first case—*The Moorcock*—is the most historically significant and concerns a term implied in fact. The court prefers a necessity to reasonableness. In *The Moorcock*,[31] the court held that the defendant was liable because there was an implied term in the contract that they would take 'reasonable care to ascertain that the bottom of the river was in such a condition as not to endanger a vessel using

---

[29] See Derry v Peek (1889) 14 App Cas 337; Heilbut Symons & Co. v Buckleton (1913) AC 30.

[30] (1971) 3 All ER 237.

[31] (1889) 14 PD 64. The defendants agreed to allow the claimant to unload his vessel at their wharf. While the vessel was moored the tide fell and the uneven conditions of the river bed damaged the ship.

the wharf in an ordinary way'. Two important points emerge from the judgment. The first is the basis for implying the term, and the second is the test for when a term will be implied. As regards the first, the Court of Appeal held that there was indeed an implied term in the contract and that the term was based on the presumed intention of the parties as well as on "reason". As regards the test, the court held that the test is based on the need to give "business efficacy" to the contract which is a test based on necessity rather than reasonableness. The question is whether "is it necessary to imply a term in order to make this contract work i.e. to give it business efficacy."

In *Shirlaw v Southern Foundries*,[32] it was said that a term could never be implied unless it was obvious to both parties so that if a bystander suggested it to the parties at the time of the contract they would both say 'yes of course that is included'. This has been called the "officious bystander test" and emphasises that the test is a high one. The question is not whether it is reasonable to imply a term but whether it is such a necessary term that both parties would have considered it obvious and necessary to the contract.

However, the idea that a term can only be implied when necessary was challenged by Lord Denning in the leading modern case *Liverpool City Council v Irwin*.[33] Lord Denning adopted reasonableness as the basis for implying terms. The defendant appealed to the House of Lords who held that there was an implied term to take reasonable care to keep the property in repair but that there had been no breach on the facts. Lord Wilberforce stated that Lord Denning's test of reasonableness was far too expansive and that necessity must always be the touchstone. He stated: "In my opinion such obligation should be read into the contract as the nature of the contract itself implicitly requires, no more, no less: a test, in other words, of necessity."

In *Liverpool City Council v Irwin*, the House of Lords introduced, *obiter dicta*, the rule that a term will be implied where it is obvious that a term has been left unstated. The obligation, then to be read into the contract, is such as the contract implicitly required, no more, no less; a test in other words of necessity. Lord Wilberforce stated:

> To construct the complete contract from these elements requires 'implications'—the supplying of what is not expressed. Not all implications are the same. Sometimes there

---

[32] (1939) 2 KB 206. The defendant was a company which appointed the plaintiff, who was then a director, to be managing director for a term of 10 years. Later, the company's Articles of Association—its constitution, in effect—were changed empowering the company to remove any director of the company, after which the plaintiff was removed. The plaintiff commenced proceedings claiming damages against the company for wrongful repudiation of the agreement.

[33] (1977) AC 239. In that case Irwin were tenants in a tower block in Liverpool. Liverpool City Council were the landlords. Irwin stopped paying rent for their maisonette and Irwin brought an action for possession and Irwin counterclaimed for the nominal figure of £10 for failing to keep common parts of the building in a state of repair. The court denied that there was any implied term to keep the common parts in repair. Irwin won at trial but lost in the Court of Appeal (Lord Denning dissenting).

may be a completed bilateral contract where the courts add terms as implied terms—in mercantile contracts where there is an established usage—the courts spell out what the parties know and would, if asked, have agreed to. Sometimes the court will add to the completed contract a term without which the contract will not work—The Moorcock. This is a strict test which may vary from time to time. Another variety of implication is that of reasonable terms—but the principle expressed by Lord Denning in the Court of Appeal goes beyond sound authority. In this case we have a fourth type or shade on the spectrum. Here the court is trying to establish what the contract is where the parties have not stated it.

In *BP Refinery v Shire of Hastings*,[34] Lord Simon, in his decision, identified five requirements for implying a specific term as such: "(1) It must be reasonable and equitable; (2) it must be necessary to give business efficacy to the contract so that no term will be implied if the contract is effective without it; (3) it must be so obvious that it 'goes without saying'; (4) it must be capable of clear expression; (5) it must not contradict any express term of the contract."

In *Investors Compensations Scheme v West Bromwich Building Society*,[35] Lord Hoffmann stated that, obiter dicta, the reason for complexity, in negotiated or committee drafted contracts, was that the people writing the contract did not know what they wanted, therefore they could not describe it clearly. In particular, where the parties wanted two inconsistent things, they would use complex language to disguise the inconsistency. In the Privy Council decision delivered, Lord Hoffmann, after making some general observations in respect of the approach of the courts, stated that a court faced with a proposed implied term simply needs to ask one question: "Is that what the instrument, read as a whole against the relevant background, would reasonably be understood to mean. The implication of the term is not an addition to the instrument. It only spells out what the instrument means."

Lord Hoffmann stated five principles as to the interpretation of terms in a contract:

(1) Interpretation is the ascertainment of the meaning which the document would convey to a reasonable person having all the background knowledge which would reasonably have been available to the parties in the situation in which they were at the time of contract. (2) The background was famously referred to by Lord Wilberforce as the 'matrix of fact', but this phrase is, if anything, an understated description of what the background may include. Subject to the requirement that it should have been reasonably available to the parties and to the exception to be mentioned next, it includes absolutely anything which would have affected the way in which the language of the document would have been understood by a reasonable man. (3) The law excludes from the admissible background the previous negotiations of the parties and their declarations of subjective intent. They are admissible only in an action for rectification. The law makes this distinction for reasons of practical policy and, in this respect only, legal interpretation differs from the way we would interpret utterances in ordinary life. The boundaries of this exception are in some respects unclear. But this is not the occasion on which to explore them. (4) The meaning which a

---

[34] BP Refinery (Westernpoint) Pty Limited v The President, Councillors and Ratepayers of Shire of Hastings (1978) 52 AUR 20. BP entered into a contract with the state government to build an oil facility with a subsidised tax. Subsequently. BP was taken over by BP Australia. Shire tried to imply a term in the contract that the preferential tax would cease to operate. .
[35] (1998) 1 All ER 98.

document (or any other utterance) would convey to a reasonable man is not the same thing as the meaning of its words. The meaning of words is a matter of dictionaries and grammars; the meaning of a document is what the parties using those words against the relevant background would reasonably have been understood to mean. The background may not merely enable the reasonable man to choose between the possible meanings of words which are ambiguous but even (as occasionally happens in ordinary life) to conclude that the parties must, for whatever reason, have used the wrong words or syntax. (5) The 'rule' that words should be given their 'natural and ordinary meaning' reflects the commonsense proposition that we do not easily accept that people have made linguistic mistakes, particularly in formal documents. On the other hand, if one would nevertheless conclude from the background that something must have gone wrong with the language, the law does not require judges to attribute to the parties an intention which they plainly could not have had.

In *Attorney General of Belize and others v Belize Telecom Ltd and Another*,[36] Belize Telecom Ltd argued that the two directors were irremovable unless they resigned, died or vacated office under article 112 of the articles of association, which provided for vacation in circumstances of conflict of interest, bankruptcy or other specified reasons. The Attorney General of Belize argued that the articles should be construed as providing by implication that a director who had been appointed by a person holding the requisite percentage of ordinary shares vacated his office if his appointer ceased to hold such a shareholding. The Court of Appeal held against the Attorney General of Belize. However, Lord Hoffmann, who delivered the advice of the Privy Council, decided the appeal should be allowed.

Lord Hoffmann, as established in his own judgment in *Investors Compensation Scheme Ltd v West Bromwich Building Society* (1998) (*supra*), recited the established principle of construction that, in discovering what an instrument means, it may be that the meaning is not necessarily or always what the authors or parties to the document would have intended. He stated: "It is the meaning which the instrument would convey to a reasonable person having all the background knowledge which would reasonably be available to the audience to whom the instrument is addressed."

2.9 Parties to Contract: Privity and Transfer of Benefits

The general concept of a contract involves obligations voluntarily undertaken towards specific people. Under law, this concept was developed restrictively into the doctrine of privity of contract, whereby a person who is not a party to a

---

[36] (2009) UKPC 10. The decision concerned the construction of the articles of association of a company incorporated to take over the undertaking of the Belize Telecommunications Authority, a public body which had been the monopoly provider of telecommunication services in Belize. The articles provided that any person who held both a 'golden share' plus 37.5% or more of the issued ordinary shares in the company could appoint or remove two directors. The articles were silent as to what was to happen to the two directors if, as happened in this case, the golden shareholder no longer held the requisite percentage of ordinary shares.

contract can neither sue on, nor rely on a defence based on that contract. The doctrine of privity means that a contract cannot, as a general rule, confer rights or impose obligations arising under it on any person except the parties to it.

The law has a number of doctrines as to the capacity of certain classes of person to enter into binding contracts. There are various doctrines as to whether contracts made by such persons are void, voidable or unenforceable by one side or the other. More significant are doctrines concerning corporations. A reference in a statute to a person includes a corporation. Partnerships are not corporations, but they may contract in the name of the partnership, on the basis of joint or joint and several liabilities of the individual partners.

Lord Diplock rationalised the rule of privity in *Dunlop v Lambert*[37] as an application of the principle:

> That in a commercial contract concerning goods where it is in the contemplation of the parties that the proprietary interests in the goods may be transferred from one owner to another after the contract has been entered into and before the breach which causes loss or damage to the goods, an original party to the contract, if such be the intention of them both, is to be treated in law as having entered into the contract for the benefit of all persons who have or may acquire an interest in the goods before they are lost or damaged, and is entitled to recover by way of damages for breach of contract the actual loss sustained by those for whose benefit the contract is entered into.

As to the principles of privity, the common law reasoned that:

1. There is the principle that consideration must move from the promisee. See *Tweddle v Atkinson*.[38]
2. Only a promisee may enforce the promise meaning that if the third party is not a promisee he is not privy to the contract. See *Dunlop Tyre Co v Selfridge* (1915) (*supra*).
3. A contract between two parties may be accompanied by a collateral contract between one of them and a third person relating to the same subject-matter. See *Shanklin Pier v Detel Products* (1951) (*supra*).

The implications of privity are significant in construction. For example, subcontractors are not parties to the main contract between employer and main contractor, so that the employer cannot sue the subcontractor for breach of contract, and vice versa. The theory is that rights and obligations should be pursued up and down the contractual chain. There must be, however, an intention to create a collateral contract before that contract can be formed. The burden of a contract cannot be transferred without the consent of the other party. If a new party is substituted with the consent of the other, it is called a novation; the new contract then refers back to the commencement of the original contract.

---

[37] (1839) 7 ER 824.

[38] (1861) 1 B&S 393. The fathers of a husband and wife agreed in writing that both should pay money to the husband, adding that the husband should have the power to sue them for the respective sums. The husband's claim against his wife's fathers' estate was dismissed, the court justifying the decision largely because no consideration moved from the husband.

The topic of assignable rights has been the subject of intense judicial and academic consideration. It pitches against each other two fundamental principles, namely, freedom of contract and freedom to dispose of one's property. The collision of these two principles is compounded by long standing difficulties of characterising assignment as to the principle that assignment is an exception to privity of contract. The benefit of a contract may be assigned to a third party without the consent of the other contracting party. If this is not desired, it is open to the parties to agree that the benefit of the contract shall not be assignable by one or either of them, either at all or without the consent of the other party. Certain expressed terms in special construed contracts conditions providing for a prohibition against assignment of obligations under the contract are not usually contrary to public policy, and a purported assignment in breach of that condition was ineffective.

In *Woodar Investment Development v Wimpey Construction*,[39] the House of Lords rejected the basis on which Lord Denning had arrived at his decision in *Jackson v Horizon Holidays Ltd*,[40] where it was held that the plaintiff could not only recover damages for himself but also the discomfort experienced by his family, and reaffirmed the view that a contracting party cannot recover damages for the loss sustained by the third party. Lord Keith of Kinkel stated that it was open to the court to declare that: "In the absence of evidence to show that he has suffered no loss, A, who has contracted for a payment to be made to C, may rely on the fact that he required the payment to be made as prima facie evidence that the promise for which he contracted was a benefit to him and that the measure of his loss in the event of non-payment is the benefit which he intended for C but which has not been received."

in *Linden Gardens Trust Ltd v Lenesta Sludge Disposals*,[41] the law on assignment was considered in depth where the Court of Appeal held that a party to a contract which contains prohibitions on assignment can transfer the commercial benefit of such a contract to a third party purchaser by declaring itself a trustee of the benefit of the contract. The effect of a particular restriction on assignment will primarily be a matter of construction; the benefits of a contract may be assigned to a third party, either by statutory assignment, which requires written notice to be given to the client; or by equitable assignment which does not require such notice.

---

[39] (1977) 1 W.L.R. 277. Wimpey agreed to buy land from Woodar for a sum of £850,000 of which £150,000 was to be paid to Transworld. A month later Wimpey sent a letter purporting to rescind the contract and Woodar sued for damages including the £150,000 payable to Transworld.

[40] (1975) 1 WLR 1468. The plaintiff entered into a contract for himself and his family. The holiday provided failed to comply with the description given by the defendants in a number of respects. The plaintiff recovered damages and the defendants appealed against the amount. Lord Denning MR thought the amount awarded was excessive compensation for the plaintiff himself, but he upheld the award on the ground that the plaintiff had made a contract for the benefit of himself and his family, and that he could recover for their loss as well as for his own.

[41] (1994) AC 85. Stock Conversion Ltd sued Lenesta Sludge to remove asbestos from a property it owned for which Lenesta sludge failed to do so appropriately. Stock Conversion Ltd assigned the lease to Linden Gardens and Linden Gardens pursued the court case against Lenesta Sludge.

It was suggested by Lord Browne-Wilkinson, in the *Linden Gardens v Lenesta Sludge* case, that restrictions should not readily be construed as precluding transfers which do not affect the original obligor under the contract but which are merely intended to operate between the assignor and assignee. Both contractor and employer were aware that the property was going to be occupied and possibly purchased by third parties. Lord Browne-Wilkinson said:

> A prohibition on assignment normally only invalidates the assignment as against the other party to the contract so as to prevent a transfer of the chose in action: in the absence of clearest words it cannot operate to invalidate the contract as between the assignor and the assignee and even then it may be ineffective on the grounds of public policy. In this case the court was able to give effect to what it saw as the intention of the parties by means of a trust.

The rule in *Dunlop v Lambert* (1839) (*supra*), as expounded by Lord Diplock, was applied in *St. Martins Property Corporation Ltd v Sir Robert McAlpine Ltd*,[42] a case arising out of breach of a building contract whereby St. Martins had contracted with McAlpine for the multi-purpose development of a site in London. After the practical completion of the works a serious defect was discovered which was remedied at a substantial cost paid for initially by St. Martins who were later reimbursed by the assignee company. The defect was alleged to have resulted from a breach of contract occurring after the assignment. St. Martins sued McAlpine who maintained that since St. Martins had suffered no loss they were only entitled to nominal damages.

In *St. Martins v McAlpine*, Lord Browne-Wilkinson concluded that St. Martins were entitled to recover substantial damages. He stated:

> In my judgment the present case falls within the rationale of the exceptions to the general rule that a plaintiff can only recover damages for his own loss. The contract was for a large development of property which, to the knowledge of both Corporation and McAlpine, was going to be occupied, and possibly purchased, by third parties and not by Corporation itself. Therefore it could be foreseen that damage caused by a breach would cause loss to a later owner and not merely to the original contracting party, Corporation. As in contracts for the carriage of goods by land, there would be no automatic vesting in the occupier or owners of the property for the time being who sustained the loss of any right of suit against McAlpine. On the contrary, McAlpine had specifically contracted that the rights of action under the building contract could not without McAlpine's consent be transferred to third parties who became owners or occupiers and might suffer loss. In such a case, it seems to me proper, as in the case of the carriage of goods by land, to treat the parties as having entered into the contract on the footing that Corporation would be entitled to enforce contractual rights for the benefit of those who suffered from defective performance but who, under the terms of the contract, could not acquire any right to hold McAlpine liable for breach. It is truly a case in which the rule provides a remedy where no other would be available to a person sustaining loss which under a rational legal system ought to be compensated by the person who has caused it.

---

[42] (1994) 1 AC 85. The contract contained a clause prohibiting the assignment of the contract by St. Martins without the consent of McAlpine. Some 17 months after the contract date St. Martins assigned to another company in the group for full value their whole interest in the property without attempting to obtain the consent of McAlpine.

In *Darlington BC v Wiltshier (Northern) Ltd*,[43] the builders took the point that the council, as assignee, had no greater rights under the contracts than the finance company had and that, as the finance company did not own the site, it had suffered no loss. The Court of Appeal applied the principles in *Linden Gardens v Lenesta Sludge* (1994) (*supra*) to enable a third party, owner of a building, to recover damages for defective work from the contractor, even though their contracts had been with a finance company which had no obligation or intention to repair. LJ Steyn held that: "A third party may sue on a contract to recover damages for defects if the benefit of a building contract was intended for them and had been assigned to him."

In *Panatown Ltd v Alfred McAlpine Construction Ltd*,[44] the main issue that was raised if "whether there is a general rule of English law that a person cannot recover substantial damages for breach of contract where he himself has suffered no loss by reason of the alleged breach." The House of Lords held that if a person enters into a contract acting as agent for a principal, but without the identity or existence of that principal being disclosed to the other party, it is known as the "undisclosed principal situation". If the other party wishes to enforce the contract after becoming aware of the existence and identity of the undisclosed principal, he has the right to bring an action against either the agent or the undisclosed principal. However, he must make an election which party to sue. The undisclosed principal may also sue directly on the contract, but his rights are subject to any rights of set-off against the agent which have accrued before the identity and existence of the principal was disclosed.

According to Lord Clyde:

> The mere fact that the building contractor, McAlpine, has entered into a separate contract in different terms with another party with regard to possible defects in the building which is the subject of the building contract cannot of itself detract from its obligations to the employer under the building contract itself. And I approach the matter as follows. First of all, it seems to me that where one party (A) is permitted by the owner of land (B) to procure the carrying out of building work on B's property, A, if he procures a builder to do the work and the work is commenced, must be under some obligation with regard to its completion. In the present case, although it was held by Judge Thornton Q.C. that there was no contractual obligation on Panatown to carry out and complete the development satisfactorily, nevertheless there was a contractual obligation on Panatown to procure a building contract. By parity of reasoning with the example I have given, it must have been

---

[43] Darlington Borough Council v Wiltshier Northern Ltd. (1995) 3 All ER 895. The council owned land on which it wanted to build a recreational centre. Construction contracts were entered into not by the council but by a finance company. The finance company then assigned to the council its rights under the building contracts, and the council claimed damages from the builders for breach of the contracts.

[44] (2000) 4 All ER 97. Panatown employed McAlpine to build a building on land owned by UIPL. The work was defective. Panatown has sought to terminate the contract on the ground of McAlpine's failure in performance. Panatown has suffered no loss. UIPL owns a defective building, which requires a significant expenditure for its repair, and has been unable for a considerable period to put the building to a profitable use. Panatown seeks to recover, from McAlpine the loss which UIPL has suffered.

implicit in that contractual obligation that, if the builder's work was defective and the defects were not rectified by him, and Panatown should in consequence recover damages from the builder for breach of contract, Panatown should instruct another builder to rectify the defects, using the damages recovered by it to finance the remedial work.

2.10 Letters of Intent

In an ideal world all contracts should be properly drawn up and signed before work commences whether the contracts are between employer and main contractor or main contractor and subcontractor. It seems that in ever increasing instances this is not the case. Fast track construction methods often overtake the procedures for drawing up the contract, which in many instances lacks essential urgency. This applies in the construction sector and can often lead to disputes which prove time consuming and expensive to resolve.

In many cases, however, the parties are anxious to make an early start even before all the key contractual matters have been agreed and long before a formal contract has been drawn up. In the absence of a formal contract it is now common practice for the employer to send the contractor a letter of intent which allows the contractor to make a start on site or to commence design or order materials.

It is of limited advantage to a contractor or subcontractor to learn that he is entitled to a payment if there is no agreement as to how much the payment will be. From this decision it can readily be seen that even if a letter of intent includes a specific instruction to undertake work it does not necessarily mean that a contract has come into being. It is common practice for the parties to sign the letter of intent but this in itself does not mean that a contract has come into being.

To establish that a contract has been concluded not only requires evidence of agreement by the parties on all the terms they consider essential, but also sufficient certainty in their dealings to satisfy the requirement of completeness. Letters of intent as traditionally drafted fail on both counts since they were usually incomplete statements preparatory to a formal contract coming into operation. Courts are often called upon to decide whether or not the wording in a particular letter of intent is sufficient to create a contract.

When the parties fall out over the wording of a letter of intent, usually in relation to payment, the court has to decide whether the wording is sufficiently explicit to form a contract. This in essence provides that if the contractor carries out work as specified in the letter of intent the employer will make a payment. This is in no way a contract in the normal sense as it usually lacks the full details as to the extent of work which is required to be undertaken and all the terms which are to apply.

Unfortunately, while the letter of intent may include an instruction to undertake work, often it does not specify the manner in which payment is to be made. The parties are fairly relaxed as they both are under the impression that a formal contract will be produced reasonably quickly containing all the necessary terms and therefore there is no reason for concern. It is only when a sticking point is reached and no contract is agreed that the parties begin to argue as to the manner and amount of payment to be made. The courts often resolve this problem by holding that payment should be made on a *quantum meruit* basis; which in essence means a sum which is merited by the work undertaken or in other word a fair and reasonable sum. This only leads to another problem being how much is a reasonable sum.

In many instances, contracting parties are more reluctant to bind themselves commercially and seem to be relying on letters of intent. The issue which arises in this case is a question of construction of the letter of intent. There has been a substantial amount of authority about letters of intent, particularly in the context of construction contracts.

In *Turriff Construction Ltd v Regalia Knitting Mills Ltd*,[45] The contractor tendered successfully for the design and construction of a factory. They sought from the employer 'an early letter of intent to cover us for the work we will now be undertaking'. They received such a letter, stating that 'the whole to be subject to agreement on an acceptable contract'. The contractor then carried out the design work, which was necessary before planning permission and estimates could be obtained for the project. Six months later the employer abandoned the project. HHJ Fay held that a letter of intent was regarded as of no contractual effect as he said: "A letter of intent is no more than the expression in writing of a party's present intention to enter into a contract at a future date. Save in exceptional circumstances, it can have no binding effect."

In the case of *British Steel Corporation v Cleveland Bridge* (1984) (*supra*), the court held, *obiter dicta* that if work is done pursuant to a request contained in a letter of intent it will not matter whether a contract did or did not come into existence because if the party who has acted on the request is simply claiming payment his claim will usually be based on a *quantum meruit*. Lord Justice Goff said:

> Now the question whether in a case such as the present any contract has come into existence must depend on the true construction of the relevant communications which have passed between the parties and the effect (if any) of their action pursuant to those communications. There can be no hard and fast answer to the question whether a letter of intent will give rise to a binding agreement; everything must depend on the circumstances of the particular case. In most cases where work is done pursuant to a request contained in a letter of intent, it will not matter whether a contract did or did not come into existence; because if the party who has acted on the request is simply claiming payment, his claim will usually be based upon a quantum meruit, and it will make no difference whether the claim is contractual or quasi-contractual. As a matter of analysis the contract (if any) which may come into existence following a letter of intent may take one of two forms— either there may be an ordinary executory contract, under which each party assumes reciprocal obligations to the other; or there may be what is sometimes called an 'if' contract, i.e a contract under which A requests B to carry out a certain performance and promises B that, if he does so, he will receive a certain performance in return [and pay]

---

[45]   (1971) 9 BLR 20.

usual remuneration for his performance. The latter transaction is really no more than a standing offer which, if acted upon before it lapses or is lawfully withdrawn, will result in a binding contract.

Sometimes the letter of intent will stipulate that certain standard conditions are to apply. The case of *Harvey Shopfitters v ADI*[46] is an example of how notwithstanding reference to the application of standard conditions a confused situation can arise. Harvey Shopfitters tendered for some refurbishment work and were sent a letter of intent which referred to the standard conditions which were to apply. The letter of intent also included the contract price and the date for completion and the significant wording: 'if for any unforeseen reason the contract should fail to proceed and be formalised then any reasonable expenditure incurred by you in connection with the above will be reimbursed on a quantum meruit basis'. No contract was ever formalised but the work was undertaken as if there was one. Harvey Shopfitters claimed that as no formal contract had been entered into, in keeping with the wording in the letter of intent they were entitled to payment on a quantum meruit basis. The court disagreed and held that the parties may not have signed a contract but proceeded as if they had.

In the case of *Ben Barratt and Son (Brickwork) Ltd v Henry Boot Management Ltd* (1995)[47] a letter of intent was sent by the main contractor Henry Boot which was signed and returned by Ben Barratt a subcontractor. The work involved brickwork on some halls of residence for the University of Manchester. Ben Barratt maintained that it commenced work on the basis of the letter of intent and that no formal contract was ever concluded. Henry Boot argued that there was a contract. The fact that the letter of intent was signed by Ben Barratt was not material to the decision. It was held that there was clear evidence that the parties intended to enter into a subcontract and no evidence to support the contention that they did not intend there to be a subcontract until the main contract was signed.

In *Jarvis Interiors Ltd v Galliard Homes Ltd*,[48] the preliminaries in the tender bills of quantities indicated that the contract will be executed as a deed under seal. In the letter of intent, Galliard wrote that it was its intention to enter into a contract with Jarvis and that 'in the event that we do not enter into a formal contract with you through no fault of Jarvis you will be reimbursed all fair and reasonable costs incurred and these will be assessed on a quantum meruit basis.' Over the following months, Jarvis carried out a substantial amount of work, but the parties were unable to agree upon terms. In the Court of Appeal, J. Lindsay, in agreeing that in fact a contract has been formed, stated: "The correct analysis of the legal situation, in my judgment, is that a contract came into existence on the terms of the Letter of Intent, either when it was acknowledged by Jarvis (24 March), or when Jarvis began work, or, at latest, when Jarvis entered onto the site at Galliard's request." He further said this:

---

[46]   Harvey Shopfitters Ltd v ADI Ltd (2004) 2 ALL ER 982.
[47]   (1995) CILL 1026.
[48]   (2000) BLR 33.

On the appeal no one has argued that there was as yet any contract between the parties [at the date of the issue of the letter of intent]. Moreover, I see the reference to "a formal contract" as only adding force to a view, to which I shall return, that, absent express agreement or necessary implication otherwise, there was to be no contract on the basis of the Preliminaries unless and until there was a "formal contrac", namely one, in the context of those Preliminaries, under seal. This last paragraph of the Letter of Intent, further, may also go some way to have put in the parties' minds that a relatively leisurely approach could, if necessary, be endured, at any rate by Jarvis, in the completion of a formal contract, notwithstanding that the work by Jarvis had actually begun on the show flats. So long as no fault could fairly be attributed to Jarvis they could always fall back on the not uncomfortable basis of a quantum meruit. The presence of the paragraph also in my view denies the usual force to be attributed to the dictum of Steyn L.J. in Trentham (G Percy) Ltd v-Archital Luxfer (1993) 1 Lloyd's RP 25 at 27 that the fact that a transaction is performed on both sides will often make it unrealistic to argue that there was no intention to enter into legal relations, at all events if the dictum is used to support the existence of some contract other than on a quantum meruit.

In *RTS Flexible Systems Ltd v Molkerei Alois Muller GmBH & Co KG (UK Productions)*,[49] the parties had entered into a letter of intent contract stating that the parties would enter into a contract within four weeks. The contract was held to have come to an end on the expiry of the four weeks with no formal contract having been concluded. The claimant proceeded with work (and was paid for part of it) and therefore according to the court "it was held that parties had entered into a further contract."

Lord Clarke stated that:

The different decisions in the courts below and the arguments in this court demonstrate the perils of beginning work without agreeing the precise basis upon which it is to be done. Whether there was a binding contract between parties and, if so, upon what terms, depended upon a consideration of what was communicated between them by words or conduct and whether that led objectively to a conclusion that they had intended to create legal relations and had agreed upon all the terms which they regarded, or the law required, as essential for the formulation of legally binding relations. Even if certain terms of economic or other significance to the parties had not been finalised, an objective appraisal of their words and conduct might lead to the conclusion that they did not intend the agreement of such terms to be a precondition to a concluded and legally binding agreement. Also, it was possible for an agreement subject to contract to become legally binding if the parties later agreed to waive that condition. In the instant case there had been unequivocal conduct on the part of both parties showing that it was agreed that the project would be carried out by RTS for the agreed price on the terms agreed, including the MF/1 terms. And they had agreed to be bound by those terms without the necessity of a formal written contract " He continued: "As is apparent from the above, after the Letter of Intent contract expired RTS continued to build the Equipment, delivered it to Müller and were

---

[49] (2008) EWHC 1087 TCC. A draft contractual agreement to install equipment in a factory, which was never executed as the work was commenced, completed and partly paid for during the negotiations, took effect as a binding contract as the essential terms had been agreed and neither party had intended agreement of the remaining terms to be a precondition to a concluded contract. Although the draft agreement contained a clause stating that the contract was not effective until it was executed, it was possible for parties to waive such a clause and, on the facts, these parties had done so.

partially paid for it. In those circumstances the court strongly inclines to concluding that the parties have entered into some contract even though such a contract cannot be spelt out by a classic analysis of the sequence of offer and acceptance.

In *Diamond Build Ltd v Clapham Park Homes Ltd*,[50] in the dangers and problems caused by letters of intents and especially not superseded but formal contract arrangements, Mr. Justice Akenhead said:

> This is yet another case which relates to a Letter of Intent on a construction project. The issues in this case revolve around whether the Letter of Intent had been superseded by a contract incorporating the JCT Intermediate Form of Building Contract, 2005 edition. There is an estoppel said to have arisen. The case illustrates the dangers posed by letters of intent which are not followed up promptly by the parties' processing of the formal contract anticipated by them at the letter of intent stage. The Claimant seeks a declaration that by the time its relationship with the Defendant was terminated the Letter of Intent had been replaced by the standard form contract.

In respect to the effectiveness of the letter of intent, Mr Justice Akenhead further stated that:

> I now turn to the construction of the Letter of Intent. The first question to consider is whether from its terms and its acknowledgment and acceptance by DB the Letter of Intent give rise to a contract in itself. I have no doubt that it did give rise to a (relatively) simple form of contract. My reasons are as follows: (a) Whilst the first paragraph merely confirms an intention to enter into a contract, the second paragraph effectively asks DB to proceed with the work. (b) There is an undertaking in effect pending the execution of a formal contract to pay for DB's reasonable costs, albeit up to a specific sum. (c) The fact in the penultimate paragraph that the undertakings given in the letter are to be "*wholly extinguished*" upon the execution of the formal contract point very strongly to those undertakings having legal and enforceable effect until the execution of the formal contract. (d) The fact that the Specification referred to in the Letter required a contract under seal demonstrates that the parties were operating with that in mind. (e) The very fact that DB was asked to (and did) sign in effect by way of acceptance the Letter of Intent points clearly to the creation of a contract based on the terms of the Letter of Intent itself. Although this is a simple contractual arrangement, it has sufficient certainty: there is a commencement date, requirement to proceed regularly and diligently, a completion date, an overall contract sum and an undertaking to pay reasonable costs in the interim.

References

Case Law

Aries Power Plant Ltd v ECE Systems Ltd (1995) 45 Con LR 111
Attorney General of Belize and others v Belize Telecom Ltd and Another (2009) UKPC 10
Barry v Heathcote Ball & Co (Commercial Auctions) Ltd (2001) 1 All ER 865
Ben Barratt and Son (Brickwork) Ltd v Henry Boot Management Ltd (1995) CILL 1026

---

[50] (2008) EWHC 1439 TCC.

Blackpool and Fylde Aero Club v Blackpool Borough Council (1990) 3 All ER 25
BP Refinery (Westernpoint) Pty Limited v The President, Councillors and Ratepayers of Shire of
    Hastings (1978) 52 AUR 20
British Steel Corporation v Cleveland Bridge & Engineering Co Ltd (1984) 1 All ER 504
Brogden v Metropolitan Railway Co. (1877) 2 App Cas 666
Carlill v Carbolic Smoke Ball Co (1892) 2 QB 484
Courtney & Fairbairn v Tolaini (Brothers) Ltd (1975) 1 All ER 716
Currie v Misa (1875) LR 10 Exch 153
Darlington Borough Council v Wiltshier Northern Ltd (1995) 3 All ER 895
Diamond Build Ltd v Clapham Park Homes Ltd (2008) EWHC 1439 TCC
Dunlop v Lambert (1839) 7 ER 824
Dunlop Pneumatic Tyre Co Ltd v Selfridge & Co Ltd (1915) AC 847
Errington v Errington & Woods (1952) 1 KB 290
Foley v Classique Coaches Ltd (1934) 2 KB 1
G. Percy Trentham Ltd v Archital Luxfer Ltd (1993) 1 Lloyd's Rep 25
Harvela Investments Ltd v Royal Trust Co. of Canada (C.I.) Ltd (1986) AC 207
Harvey Shopfitters Ltd v ADI Ltd (2004) 2 ALL ER 982
Hillas & Co Ltd v Arcos Ltd (1932) 147 LT 503
Hyde v Wrench (1840) 3 Beav 334
Investors Compensation Scheme v West Bromwich Building Society (1998) 1 All ER 98
Jackson v Horizon Holidays Ltd (1975) 1 WLR 1468
Jarvis Interiors Ltd v Galliard Homes Ltd (2000) BLR 33
Linden Gardens Trust Ltd v Lenesta Sludge Disposals (1994) AC 85
Liverpool City Council v Irwin (1977) AC 239
Nicolene Ltd v Simmonds (1953) 1 Q.B. 543
Pagnan SpA v Feed Products (1987) 2 Lloyd's Rep 601
Panatown Ltd v McAlpine Construction Ltd (2000) 4 All ER
Perry v Suffields (1916) 2 Ch 187
Port Sudan Cotton Co v Govindaswamy Chettiar (1977) 2 Lloyds Rep 5
Prenn v Simmonds (1971) 3 All ER237
RTS Flexible Systems Ltd v Molkerei Alois Muller GmBH & Co KG (UK Productions) (2008)
    EWHC 1087 TCC
St. Martins Property Corporation Ltd v Sir Robert McAlpine Ltd (1994) 1 AC 85
Shanklin Pier Ltd v Detel Products Ltd (1951) 2 All ER 471
Shirlaw v Southern Foundries (1939) 2 KB 206
The Aramis (Cargo Owners) v Aramis (Owners) (1989) 1 Lloyd's Rep 213
The Elli 2, Ilyssia Compania Naviera SA v Bamaodah, (1985) 1 Ll R 107
'The Moorcock' (1889) 14 PD 64
Trollope & Colls v Atomic Power Constructions Ltd (1962) 3 All ER 1035
Turriff Construction Ltd v Regalia Knitting Mills Ltd (1971) 9 BLR 20
Tweddle v Atkinson (1861) 1B & S 393
Williams v Roffey Bros & Nicholls (Contractors) Ltd (1991) 1 QB 1
Woodar Investment Development Ltd v Wimpey Construction U.K. Ltd (1977) 1 W.L.R. 277

Books

Adams J, Brownsword R (2007) Understanding contract law. Sweet & Maxwell, UK
Adams J, Brownsword R (1995) Key issues in contract. Butterworths, London
Allen R, Martin S (2010) Construction law handbook. Aspen Publishers, Maryland
Beatson J, Burrows A, Cartwright A (2010) Anson's law of contract. Oxford University Press,
    Oxford
Beale H (2010) Chitty on contract. Sweet & Maxwell Ltd, UK

Beale H, Bishop W, Furmston M (2008) Contract: cases and materials. Oxford University Press, Oxford

Bockrath J, Plotnick F (2010) Contracts and the legal environment for engineers and architects. McGraw-Hill, USA

Chen-Wishart M (2010) Contract law. Oxford University Press, Oxford

Furmston M, Cheshire GF, Fifoot CHS (2006) Cheshire, fifoot and furmston's law of contract. Oxford University Press, Oxford

Halson (2001) Contract law. Longman, London

Koffman L, Macdonald E (2007) The Law of contract. Oxford University Press, Oxford

McKendrick (2009) Contract law. Palgrave Macmillan, Basington, Hampshire

O'Sullivan J, Hilliard J (2006) The law of contract. Oxford University Press, Oxford

Peel E (2007) Treitel on the law of contract. Sweet & Maxwell, UK

Poole J (2010a) Textbook on contract law. Oxford University Press, Oxford

Poole J (2010b) Casebook on contract law. Oxford University Press, Oxford

Richards P (2009) Law of contract. Pearson Longman, Essex

Treitel G (2004) An outline of the law of contract. Oxford University Press, Oxford

Wallace D (1995) Hudson's building and engineering contract. Sweet & Maxwell, UK

White N (2008) Construction law for managers,architects and engineers. Delmar Cengage Learning, USA

Chapter 3
Construction Contracts: Obligations, Vitiations and Remedies

3.1 Contractual Risks

Some of the provisions in certain contracts' general conditions not only expressly allow to abuse the rules in regard to fairness and reasonableness, the causes and situations when a contractor should claim, how to claim and when to claim, they positively encourage it. In particular, the contractor is encouraged to exaggerate the time effects of claim events because that lead to greater entitlement for payment, that is, the reverse of what should be encouraged.

The trend in construction contracts has been shifting in recent years to include more risk-sharing provisions and to reallocate increasing numbers and proportions of risks onto the employer. Some forms of contracts probably place even more risk on the employer, but these contracts made big advances by dealing with time and money, and sharing responsibilities and risks in an integrated way. This helps to reduce abuse, but it does place higher demands particularly on the engineer and the architect to address compensation events due to changed conditions promptly and to act fairly and with due diligence when certifying payments.

There is a recent tendency for the courts to apply what might be called the *portia* approach, favouring strongly interpretations of contractual provisions which presume that the parties did not intend a gamble and, therefore, imposing a presumption of business common sense; in other words, courts are discouraging the idea of first try to get out of the clause, but next try to construe things to your advantage.

Sharing, structuring and minimising the impacts of risks can be summarised by the following:

1. First, there is a distinction between risks which can be considered as insurable and those which cannot. With insurable risks the object of the contract is to ensure that the risks are allocated entirely to the party required to be insured. Since insurance is generally on an indemnity basis; the insured must be liable before the insurers will pay.

A. D. Haidar, *Global Claims in Construction*,
DOI: 10.1007/978-0-85729-730-3_3, © Springer-Verlag London Limited 2011

2. Second, the various criteria suggested for deciding risk allocation policy are neither necessarily consistent nor readily compatible. For example, the extent of unforeseeable ground conditions is best controlled by the employer though pre-contract ground investigations, but the consequences of unforeseen ground conditions, when encountered, are best managed by the contractor.
3. Third, structuring contracts, for risk sharing rather than straightforward risk allocation, tends to open up the scope for abuse.

The principal contractual provisions dealing with allocation of non-insurable risks are categorised under a number of headings:

- Quantities. These relate to the difference between actual quantities executed and quantities in the bill of quantity.
- Engineer's instructions and standards. These relate to the bidding and the execution processes.
- Changes in laws, regulations and taxes.
- Unforeseen physical conditions including soil conditions, weather and existing and neighbouring infrastructure facilities.
- Impossibility and *force majeure*.

The huge matrix of requirements, constraints and responsibilities, in order to deliver the complex large projects, will need to be project managed and programmed by applying the effective management of the numerous interrelated variables on a general scale. The contractor's unique expertise coupled with his local knowledge and his historical endeavour in consistently delivering the type of projects he is accustomed to deliver even in the most challenging environments will place him usually at the forefront of carrying the usual risks associated with the works. In summary, the contractor is responsible for achieving the following objectives:

1. To integrate the construction program for all the activities due to be instructed including the main works, the logistics, permits and resources.
2. To apply the best modern management techniques to achieve the predetermined objectives of achieving aims at cost, time and purpose built quality.
3. To provide alternative solutions for the construction and fabrication considering alternative plans in terms of LEED,[1] sustainability and green technology.
4. To provide alternative solutions in order to assure that the development scheme will be a successful scheme in terms of construction methods, re-usage of temporary facilities and infrastructure works such as transportation, electricity and water resources.

The contractor, in order to make the project successful, optimise costs and achieve risks that are minimised, must also implement the following methodologies and steps:

---

[1] Leadership in Energy & Environmental Design.

1. Project scope management to ensure that all the works required are to be completed according to the program in place and according to the costs envisaged.
2. Project time management to provide an effective project schedule.
3. Project cost management to identify needed resources and maintain budget control.
4. Project quality management to ensure functional requirements are met.
5. Project human resource management for the development and effective employment of project key personnel including government committees and managers.

The employer and the contractor have a duty to express contract provisions that are fair and reasonable without avoiding the duty of care responsibilities imposed on both of them. There has been an unspoken assumption that the contractor should be left to manage the performance of the works based on the design available and that certain categories of events are regarded as giving the contractor entitlement to extension of time for completion, but no financial compensation. This factor is, however, affected by certain mitigated factors in order to subdue the effects of these delays. Trying to shift the risk from one party to the other or in case one party tries to impose its rules, regulations and procedures on another can be disfavoured by the courts especially if certain doctrines apply which will be discussed in this chapter.

3.2 Professional Obligations in Construction

It is an accepted point, in fact and in law, that the contractor comes into a relationship with the engineer which is the result of the contractor entering into the contract with the employer and of the engineer having been engaged by agreement with the employer to perform the functions required under the contract. The engineer assumes the obligation, under its agreement with the employer, to act fairly and impartially in performing its functions. The engineer is under a contractual duty to the employer to act with proper care and skill.

The contract provides for the correction, by the process of arbitration or the courts, of any error on the part of the engineer and if there is any real scope for an error on the part of the engineer which would not be at once detected by the contractor. The court will, at least in the absence of any factual basis for the engineer to have foreseen any other outcome, proceed on the basis that the contractor would recover the sums which it ought to recover under the contract. It is, foreseeable that a contractor under such an arrangement may suffer loss by being deprived of prompt payment as a result of negligent under-certification or negligent failure to certify by the engineer and in the case, among others, that the engineer does not provide an extension of time due to unforeseen conditions for which the contractor is not liable.

The contractual duty of the engineer to act fairly and impartially, owed to the employer, is a duty in the performance of which the employer has a real interest. If the engineer should act unfair to the detriment of the contractor, claims will be made by the contractor to get the wrong decisions put right. If court proceedings are necessary, then the employer will be exposed to the risk of costs in addition to being ordered to pay the sums which the engineer should have allowed. If the decisions and the advices of the engineer, which caused the proceedings to be taken, were shown by the employer to have been made and given by the engineer in breach of the engineer's contractual duty to the employer, the employer would recover its losses from the engineer. There is, therefore, not only an interest on the part of the employer, in the due performance by the engineer to act fairly and impartially, but also a sanction which would operate, in addition to the engineer's sense of professional obligation, to deter the engineer from the careless making of unfair or unsustainable decisions adverse to the contractor.

The respective rights of the parties should be of such a nature that they might be fairly enforced whatever contingencies might arise and that, if such conditions were adopted, it should be understood by all the parties that in the event of a dispute arising every clause would be enforced without question. It might also be observed that the parties obligations in their contractual arrangement have been based on the principle that the design of the permanent works is, generally, carried out by someone other than the contractor.

The central question which arises is that the contractual structure of the contract into which the contractor was prepared to enter with the employer implies that the contractor will look to the engineer by way of reliance for the proper execution of the latter's duties under the contract. In other words, although the parties were brought into close proximity in relation to the contract, a failure by the engineer or the architect to carry out their duties under the contract would foreseeably cause loss to the contractor which was not properly recoverable under its rights against the employer. It is immediately apparent that there is no simple unqualified answer to the question 'Does the engineer owe a duty to the contractor to exercise reasonable skill and care?' but that this question can only be answered in the context of the factual matrix including especially the contractual structure against which such duty is said to arise. This creates the complex issue of matrix of duties of care and the doctrines of privity and agency.

In *Ranger v Great Western Railway Company*,[2] where Ranger was the contractor engaged on the construction of works including, *inter alia*, the Avon Bridge and the engineer was Brunel. Ranger's bills asserted fraud on the part of the company through their engineer in two relevant respects. The first was that inspection pits misled the contractor into underestimating the hardness of the rock to be excavated in the tunnel and he was thereby induced to tender an uneconomically low price. The second point, of greater subtlety, was that unbeknown to

---

[2] (1854) 5 HL Cas 72.

the contractor Brunel was a shareholder in the company. Lord Cranworth, in affirming the duties of an engineer, said:

> It is not necessary to state the duties of the engineer in detail: he was, in truth, made the absolute judge, during the progress of the works, of the mode in which the appellant was discharging his duties; he was to decide how much of the contract price of £63,028 from time to time had become payable; and how much was due for extra works; and from his decision, so far, there was no appeal. The contention now made by the appellant is, that the duties thus confided to the principal engineer were of a judicial nature; that Mr. Brunel was the principal engineer by whom those duties were to be performed, and that he was himself a shareholder in the company; that he was thus made a judge, or arbitrator, in what was, in effect, his own cause.

The case of *Sutcliffe v Thackrah*[3] established that an architect owes a duty of care towards his client and the contractor, in the performance of all duties including contract administration and certification, and could be liable for negligence in the performance of those duties. The House of Lords acknowledged, *obiter dicta*, that a professional consultant had an implied duty to act impartially when deciding questions between its client and the contractor; this means acting independently, honestly, fairly and without bias. Lord Salmon stated:

> No one denies that the architect owes a duty to his client to use proper care and skill in supervising the work and in protecting his client's interests. That, indeed, is what he is paid to do. Nevertheless, it is suggested that because, in issuing the certificates, he must act fairly and impartially as between his client and the contractor, he is immune from being sued by his client if, owing to his negligent supervision (or as in the present case) other negligent conduct, he issues a certificate for far more than the proper amount, and thereby causes his client a serious loss.

In *Sutcliffe v Thackrah*, Lord Reid, by emphasising the duty of an architect to act reasonably and fairly and above all in a professional manner, said:

> The employer and the contractor make their contract on the understanding that in all matters where the architect has to apply his professional skill he will act in a fair and unbiased manner in applying the terms of the contract. An architect is not an arbitrator but he has two different types of function to perform. In many matters he is bound to act on his client's instructions, whether he agrees with them or not; but in many other matters requiring professional skill he must form and act on his own opinion.

In *Arenson v Casson Beckman Rutley & Co.*,[4] Lord Salmon, in dismissing a submission made that there should not be a duty owed by the architect to the contractor since it would put the former in the risk of being liable for both the client and the contractor simultaneously, said in favouring the principles set by *Sutcliffe v Thackrach* (1974) (*supra*):

> In spite of the remarkable skill with which this argument was developed, I cannot accept it. Were it sound, it would be just as relevant in Sutcliffe v Thackrah as in the present case. The architect owed a duty to his client, the building owner, arising out of the contract

---

[3] (1974) AC 727.
[4] (1975) 3 All ER 901.

between them to use reasonable care in issuing his certificates. He also, however, owed a similar duty of care to the contractor arising out of their proximity. In Sutcliffe v Thackrah the architect negligently certified that more money was due than was in fact due; and he was successfully sued for the damage which this had caused his client. He might, however, have negligently certified less money was payable than was in fact due and thereby starved the contractor of money. In a trade in which cash flow is especially important, this might have caused the contractor serious damage for which the architect could have been successfully sued. He was thus exposed to the dual risk of being sued in negligence but this House unanimously held that he enjoyed no immunity from suit.

In *Anns v Merton London Borough Council*,[5] Lord Wilberforce, after reviewing the trilogy of cases namely, *Donoghue v Stevenson*,[6] Hedley Byrne & Co. Ltd. v Heller & Partners Ltd.,[7] and *Dorset Yacht Co. Ltd., v Home Office*,[8] introduced a two stage test for imposing a duty of care by stating the following:

> Rather the question has to be approached in two stages. First one has to ask whether, as between the alleged wrongdoer and the person who has suffered damage there is a sufficient relationship of proximity or neighbourhood such that, in the reasonable contemplation of the former, carelessness on his part may be likely to cause damage to the latter-in which case a prima facie duty of care arises. Secondly, if the first question is answered affirmatively, it is necessary to consider whether there are any considerations which ought to negative, or to reduce or limit the scope of the duty or the class of person to whom it is owed or the damages to which a breach of it may give rise.

In *Governors of the Peabody Donation Fund v Sir Lindsay Parkinson & Co. Ltd.*,[9] the court held that the true question to find negligence was whether the particular defendant owed the particular plaintiff a duty of care having the scope pleaded and that it was reasonable for that duty to be imposed. It was not reasonable to impose a duty on the local authority, in this case, to indemnify the builders from relying upon the advice of their own architects and contractors. Lord Keith stated the following:

> The true question in each case is whether the particular defendant owed to the particular plaintiff a duty of care having the scope which is contended for, and whether he was in breach of that duty with consequent loss to the plaintiff. A relationship of proximity in Lord Atkin's sense must exist before any duty of care can arise, but the scope of the duty must depend on all the circumstances of the case.

---

[5] (1978) AC 728. The claimants were tenants in a block of flats. The flats suffered from structural defects due to inadequate foundations which were 2ft 6in deep instead of 3ft deep as required. The defendant Council was responsible for inspecting the foundations during the construction of the flats. The House of Lords held that the defendant did owe a duty of care to ensure the foundations were of the correct depth.

[6] (1932) AC 562.

[7] (1964) AC 465.

[8] (1970) AC 1004.

[9] (1985) AC 210. The architects proposed a system of flexible drains for a site, but the contractors persuaded them to accept rigid drains which once laid proved inadequate at considerable cost. The local authority permitted the departure from the plans.

3.3 Duty of Care and Negligent Statements

Duty of care, stated simply, means that one must take reasonable steps to ensure that his actions do not knowingly cause harm to another individual. In such cases, the courts look to the nature of the relationship between the parties and whether the incident resulting in harm was reasonably foreseeable. It is also imperative that there is proximity or causal connection between one person's conduct and the other person's injury. If the actions of a person are not made with watchfulness, attention, caution and prudence then their actions are considered negligent; consequently the resulting damages may be claimed as negligence in a lawsuit.

Negligence is a failure to take reasonable care for the safety or well-being of others. Negligent actions are not an exercise in perfection but rather address issues of reasonableness or, put simply, what a reasonable person might have done or not done in the circumstances of a particular case. The law of negligence entitles a person to receive compensation, for loss or damage, as a result. In general terms, negligence can be established if:

- The defendant owed them a duty to take reasonable care;
- The defendant breached that duty;
- The defendant's breach of duty caused the injury or damage suffered by the plaintiff; and
- The injury or damage was not too remote a consequence of the breach of duty.

Historically, the accuracy of the statement would be warranted as a term of the contract. If the statement was incorporated in the contract there could be liability for breach of contract. Financial relief was only available if the statement was incorporated in the contract or had been made dishonestly. Such a term would be classified as a 'promissory representation' or simply a misdescription.

The law was very reluctant to allow a person to claim damages for losses suffered as a result of a statement being untrue, unless the maker of the statement had made it as part of a contract or had made the statement fraudulently. Compensatory remedies were only available in respect of misrepresentations incorporated in a contract, in which case damages would be recoverable for breach of warranty, or in respect of losses suffered as a result of a fraudulent statement, in which case damages were recoverable in the tort of deceit.

Hedley Byrne & Co. Ltd. v Heller & Partners Ltd. (1964) *(supra)*[10] was a watershed in regard to the doctrines of reasonableness, duty of care and negligent misstatement. A feature of *Hedley Byrne v Heller & Partners* is that there was an

---

[10] Hedley was advertising agent who had provided a substantial amount of advertising on credit for Easipower. Hedley became concerned that Easipower would not be in a financial position to pay the debt and sought assurances from Easipower's bank that Easipower was in a position to pay for the additional advertising which Hedley may give them on credit. The respondents, who were Easipower's bankers, gave a favourable report of Easipowers financial position. On the strength of the report given by the respondents, Hedley placed additional orders on behalf of Easipower which eventually resulted in a loss of £17,000.

approach, made to the defendant bank by or on behalf of the plaintiffs, inviting the bank to provide a service of advice and information directly to them. Lord Pearce held that there could be an implied duty of care imposed on professionals acting on behalf of employers in giving statements and advice. In this regard, he stated the following: "A duty of care created by special relationships which, though not fiduciary, gives rise to an assumption that care as well as honesty is demanded."

Further, as to the issue of duty of care imposed on a professional providing a statement, Lord Reid said:

> A reasonable man, knowing that he was being trusted or that his skill and judgment were being relied on, would, I think, have three courses open to him. He could keep silent or decline to give the information or advice sought: or he could give an answer with a clear qualification that he accepted no responsibility for it or that it was given without that reflection or inquiry which a careful answer would require: or he could simply answer without any such qualification. If he chooses to adopt the last course he must, I think, be held to have accepted some responsibility for his answer being given carefully, or to have accepted a relationship with the inquirer which requires him to exercise such care as the circumstances require.

In *Hedley Byrne v Heller & Partners*, Lord Morris, as to when men of special skills have a duty of care implied in their statements, further stated:

> My lords, I consider that it follows and that it should now be regarded as settled that if someone possessed of a special skill undertakes, quite irrespective of contract, to apply that skill for the assistance of another person who relies on such skill, a duty of care will arise. The fact that the service is to be given by means of or by the instrumentality of words can make no difference. Furthermore if, in a sphere in which a person is so placed that others could reasonably rely on his judgment or his skill or on his ability to make careful inquiry, a person takes it on himself to give information or advice to, or allows his information or advice to be passed on to, another person who, as he knows or should know, will place reliance on it, then a duty of care will arise.

Since *Hedley Byrne v Heller & Partners* (1964) (*supra*), both legislative and judicial law making constituted moves to allow, either for the first time or to a much greater extent than before, actions for damages to compensate for loss resulting from reliance on a false statement made by the defendant, without that statement being part of a contract between the plaintiff and the defendant, and without the need to prove fraud. If a contract contains a term in the contract and the term is unfair, then the term would exclude or restrict the rights of the parties to any liability to which a party to a contract may be subject by reason of any misrepresentation made by him before the contract was made or any remedy available to another party to the contract by reason of such a misrepresentation. This term shall be of no effect except insofar as it satisfies the requirement of reasonableness; and it is for those claiming that the term satisfies the requirement to show that it does.

In *Dorset Yacht Co. Ltd. v Home Office* (1970) (*supra*), Lord Morris, after observing what Lord Atkin said in *Donoghue v Stevenson* (1932) (*supra*) that it was advantageous if the law "*is in accordance with sound common sense*" and expressing the view that a special relation existed between the parties which gave rise to a duty on the former to control their duties so as to prevent them doing damage, continued himself:

> Apart from this I would conclude that in the situation stipulated in the present case it would not only be fair and reasonable that a duty of care should exist but that it would be contrary to the fitness of things were it not so. If the test whether in some particular situation a duty of care arises may in some cases have to be whether it is fair and reasonable that it should so arise the court must not shrink from being the arbiter. So in determining whether or not a duty of care of particular scope was incumbent on a defendant it is material to take into consideration whether it is just and reasonable that it should be so.

In *Esso Petroleum Co. Ltd. v Mardon*,[11] Mardon was told by an Esso employer that the estimated throughput of the Eastbank Street site would amount to 200,000 gallons a year. Mardon indicated that he thought 100,000 to 150,000 gallons would be a more realistic estimate, but he was convinced by the far greater expertise of the Esso expert and based on his statement proceeded with the lease contract. LJ Shaw said:

> It is difficult to see why, in principle, a right to claim damages for negligent misrepresentation which has arisen in favour of a party to a negotiation should not survive the event of the making of a contract as the outcome of that negotiation. It may, of course, be that the contract ultimately made shows either expressly or by implication that, once it has been entered into, the rights and liabilities of the parties are to be those and only those which have their origin in the contract itself. In any other case there is no valid argument, apart from legal technicality, for the proposition that a subsequent contract vitiates a cause of action in negligence which had previously arisen in the course of negotiation. In the present case the proposition would not save Esso from liability if they be held to have given a warranty. Thus Mr. Mardon is entitled in my view to damages for breach of warranty or for negligent misrepresentation.

3.4 Contra Proferentem Rule

The doctrine, as a general rule of construction, is stated in the Latin maxim '*verba chartarum fortius accipiuntur contra proferentem*'. As explained in Anson's Law of Contract, the rule is based on the principle "that a person is responsible for ambiguities in its own expression, and has no right to induce another to contract with it on the supposition that the words mean one thing, and then to argue for a construction by which they would mean another thing more to its advantage."

The *contra proferentem* doctrine is a deed or representation, to be construed more strongly against the person putting forward the document, whose purpose is

---

[11] (1976) 2 All ER 5. Esso Petroleum wanted an outlet for their petrol in Southport. They estimated that the throughput of petrol would reach 200,000 gallons a year by the second year after development. In addition, they would get a substantial rental from a tenant. Esso had thought that they could have the forecourt and pumps fronting on to the busy main street. But the planning authority, insisted that the station should be built 'back to front' and that is what happened.

to prevent the use of unintelligible terms through the threat of applying an interpretation in favour, not of whoever is responsible for creating such unintelligibility, but of the other party *ex ante* will. However, where both parties have been involved in agreeing the terms of a document, the courts will be reluctant to apply the *contra proferentem* doctrine; the doctrine only applies as a general rule of construction when all others have failed.

The *contra proferentem* rule is thus clearly one of default and is applied if, and only if, there is no particular condition that provides for the issue at stake; but it is also a rule which penalises the client or engineer use of unintelligible industry common terms and conditions as any doubts are always resolved against them and in favour of the contractor. This leads to an outcome that is clearly opposed to what the author of the terms would have wanted and thus acts as an incentive to word terms in a clear way, i.e., to reveal information to both the other party and the courts.

In summary, those drafting or amending contracts need to ensure that they incorporate terms which are concise, unambiguous and preferably in plain and intelligible language. Otherwise, if there is doubt about the meaning of a written term, they may find that an interpretation more favourable to the other party is applied, in appropriate circumstance. This may be contrary to their intentions, and in certain circumstances, could be financially disastrous.

In *Stevenson v Reliance Petroleum Ltd.*,[12] J. Cartwright opined: "The rule expressed in the maxim, *verba fortius accipiuntur contra proferentem*, was pressed upon us in argument, but resort is to be had to this rule only when all other rules of construction fail to enable the court of construction to ascertain the meaning of a document."

In *Peak Construction (Liverpool) v McKinney Foundations*,[13] the court held that the *contra proferentem* rule applied to extension of time clauses and liquidated damages. Part of Peak Construction claim had been for the sum of £4,205 being liquidated damages under the main contract which the employer was alleged to have been entitled to as a consequence of the overall delay. This portion of the claim was disallowed on appeal on the basis that Peak Construction would not have been liable under the main contract to pay any liquidated damages to the employer since the extension of time clause was to be construed strictly *contra proferentem* against the employer. In other words, to the extent that there was any ambiguity in regard to the extension of time clause where delay had been caused by default of the employer, this would be interpreted against the employer and the benefit of the doubt would be given given to the contractor.

In *Peak Construction v McKinney Foundations*, the court held that if the employer was in any way responsible for the failure to achieve the completion date, the employer could recover no liquidated damages whatsoever and would be left to prove such general damages as it may have suffered. In any event, the

---

[12] (1956) S.C.R. 936.
[13] (1970) 1 BLR 114.

extension of time clause in the contract provided only a limited basis for extending the time for completion in respect of the actual phrase 'delays caused by unavoidable circumstances'. This falls under the doctrine of *contra proferentem* as it is not wide enough to embrace delays due to employer's own breach, and as a consequence time had effectively become at large, and the contractor's only obligation was to complete the work within a reasonable time.

In *Horne Coupar v Velletta & Company*,[14] J. Romilly, applied the *contra proterentem* doctrine, in coming to his decision. Specifically, he reasoned as follows:

> Contra proferentem is a rule of contractual interpretation which provides that an ambiguous term will be construed against the party responsible for its inclusion in the contract. This interpretation will therefore favour the party who did not draft the term presumably because that party is not responsible for the ambiguity therein and should not be made to suffer for it. This rule endeavours to encourage the drafter to be as clear as possible when crafting an agreement upon which the parties will rely. This rule also encourages a party drafting a contract to turn their mind to foreseeable contingencies as failure to do so will result in terms being construed against them. That there is ambiguity in the contract is a requisite of the application of this rule, however, once ambiguity is established, the rule is fairly straightforward in application.

3.5 Misrepresentation

Under contract law, a misrepresentation is defined as a false statement of a material fact made by one party to another which had induced the other party to enter into a contract. Even if the statement was one of opinion, an action for misrepresentation may also happen if it can be proved that the maker of the statement did not actually believe in the truth of the opinion or if it can be established that a reasonable man having the maker's knowledge could not have honestly held such an opinion.

Misrepresentation is divided by the following categories:

- Innocent misrepresentation. Innocent misrepresentation describes a situation where the person making the statement can show he had reasonable grounds to believe his statement was true.
- Negligent misrepresentation. Negligent misrepresentation describes a statement which is made carelessly or without reasonable grounds for believing its truth.
- Fraudulent misrepresentation. Fraudulent misrepresentation occurs when a false statement is made knowingly, or without a belief that it is true or recklessly as to its truth.

In respect to fraudulent misrepresentation, the following situations are tested by the courts:

---

[14] (2010) BCSC 483.

1. There is no fraud if the person making the statement honestly believes the statement to be true.
2. A claimant will need to prove an absence of honest belief in the truth of the statement for an action for fraudulent misrepresentation to succeed.
3. It is enough to show that the person making the statement suspected it might be inaccurate, or that he neglected to make enquiries, without proving that the maker knew his statement was false.
4. Absence of reasonable grounds for a belief does not amount to fraud but may be used as evidence from which an inference can be drawn that there was no honest belief in the truth of the statement.
5. The test of misrepresentation is usually objective. Where the representation is claimed to be fraudulent, the court will inquire into the subjective state of mind of the maker of the statement.

The difficult question is whether, when an employer possesses relevant information, non-disclosure can amount to a misrepresentation. The answer depends on whether there is a duty to disclose the information. For example, insurance provisions in construction contracts are subject to a doctrine of utmost good faith, or *uberrimae fides*, which imposes a positive duty to disclose all material facts. There may be other situations where a party comes under an affirmative duty, but in relation to information on site and ground conditions, there is no general rule.

In addition, to be able to sue for misrepresentation, the false statement must have induced the formation of the contract. However, the availability and effectiveness of the remedy was very limited. At common law, rescission would only be granted if it was possible to restore the parties precisely to their original positions.

The remedies afforded by the courts in case of misrepresentation are summarised as such:

1. Where a person has entered into a contract after a misrepresentation has been made to him by another party and as a result thereof he has suffered loss, then, the person making the misrepresentation would be liable to damages, unless he proves that he had reasonable ground to believe and did believe up to the time the contract was made that the facts represented were true.
2. Where a person has entered into a contract after a misrepresentation has been made to him otherwise than fraudulently, and he would be entitled by reason of the misrepresentation to rescind the contract, then the court may declare the contract subsisting and award damages in lieu of rescission.
3. The misrepresentation need not be the sole factor that induces the formation of the contract. Silence may amount to misrepresentation if a half truth is offered or if the maker realises the statement is not true before the contract is made. Rescission was considered an appropriate remedy because the false statement of fact could not be restored by payment of money.

In *Attwood v Small*,[15] a preliminary agreement was made between the parties whereby the claimant agreed to purchase subject to being satisfied that the reports and accounts given by the defendant were accurate. It then transpired that the accounts had greatly exaggerated the income generated by the estate and the claimant sought to rescind the contract based on the misrepresentations contained in the reports and accounts. The House of Lords held that the purchaser had relied on the report of other professionals as well and so could not sue for misrepresentation and that the purchasers' application to rescind the contract on the grounds of misrepresentation must fail.

In *Attwood v Small*, Lord Brougham stated:

> It must be shown that the attempt was made, and made with success, cum fructu. The party must not only have been minded to overreach, but must actually have overreached. He must not only have given instructions to the agent to deceive, but the agent must, in fulfilment of his directions, have made a misrepresentation; and moreover, the representation so made must have had the effect of deceiving the purchaser; and moreover, the purchaser must have trusted to that representation, and not to his own acumen, not to his own perspicuity, not to inquiries of his own.

In *Derry v Peek*,[16] the court held, *obiter dicta*, that false statements not made fraudulently were classified as innocent; it was immaterial whether the statement had been made carelessly, so long as it had been made honestly and a misrepresentation is fraudulent if the maker knew it was false or did not believe in the truth of the statement or was recklessly careless whether the statement was true or false. The following statement is taken from Lord Herschell's speech:

> I think the authorities establish the following propositions: First, in order to sustain an action of deceit, there must be proof of fraud, and nothing short of that will suffice. Secondly, fraud is proved when it is shown that a false representation has been made (1) knowingly, or (2) without belief in its truth, or (3) recklessly, whether it be true or false. Although I have treated the second and third as distinct cases, I think the third is but an instance of the second, for one who makes a statement can have no real belief in the truth of what he states. To prevent a false statement being fraudulent, there must, I think, always be an honest belief in its truth. And this probably covers the whole ground, for one who knowingly alleges that which is false obviously has no such belief. Thirdly, if fraud be proved, the motive of the person guilty of it is immaterial. It matters not that there was no intention to cheat or injure the person to whom the statement was made.

---

[15] (1838) 6 Cl & Flyn 232. The claimants purchased Corngreaves estate from the defendant for £600,000. Corngreaves estate consisted of mining land, iron works and various properties including a mansion house. Many of the properties were subject to leasehold and generated income. The mines were to be worked by and profit to go to the claimant.

[16] (1889) All ER 1. The plaintiff purchased shares in the railway company based on the information contained in the prospectus. The railway company then failed to obtain the necessary government approval for steam power and the company dissolved. The plaintiff then brought an action in deceit against the railway company for fraudulent misrepresentation.

In *Bradford Building Society v Borders*,[17] Viscount Maugham, by establishing the steps required for the test of misrepresentation which he has equated with deceit, stated the following:

> My Lords, we are dealing here with a common law action of deceit, which requires four things to be established. First, there must be a representation of fact made by words, or, it may be, by conduct. The phrase will include a case where the defendant has manifestly approved and adopted a representation made by some third person. On the other hand, mere silence, however morally wrong, will not support an action of deceit. Secondly, the representation must be made with a knowledge that it is false. It must be wilfully false, or at least made in the absence of any genuine belief that it is true. Thirdly, it must be made with the intention that it should be acted upon by the plaintiff, or by a class of persons which will include the plaintiff, in the manner which resulted in damage to him. If however, fraud be established, it is immaterial that there was no intention to cheat or injure the person to whom the false statement was made. Fourthly, it must be proved that the plaintiff has acted upon the false statement and has sustained damage by so doing. I am not, of course, attempting to make a complete statement of the law of deceit, but only to state the main facts which a plaintiff must establish.

In *Museprime Properties v Adhill Properties*,[18] the court held that there will only be an inducement by misrepresentation if the statement made is material. It must represent a fact upon which a party decides to enter into the contract; although it does not have to be the sole inducement, it is sufficient that it is one of the inducements. According to the court, the rule, however, is not strictly objective. If the misrepresentation would have induced a reasonable person to enter into a contract, then the court will presume that the plaintiff was so induced, and the onus will be on the defendant to show that the claimant did not rely on the misrepresentation either wholly or in part. If, however, the misrepresentation would not have induced a reasonable person to contract, the onus will be on the misrepresentee to show that the misrepresentation induced her to act as he did.

3.6 Mistake

A mistake in law, to be operative must be of fact and not of law. It allows the parties to rescind a contract, which effectively is to put the parties back into the positions they held before the contract was made. The doctrine of mistake applies

---

[17] Bradford Third Equitable Benefit Building Society v Borders (1941) 2 ALL ER 205. This was a case in which a building society financier was not held liable for a false statement, that the house purchased had been well built, made by a group of builder-developers with whom the society had signed a contract to finance purchases of the land and houses.

[18] (1990) 36 EG 114. In a sale by auction of three properties the particulars wrongly represented the rents from the properties as being open to negotiation. The statements in the auction particulars and made later by the auctioneer misrepresented the position with regard to rent reviews. In fact, on two of the three properties rent reviews had been triggered and new rents agreed. The plaintiff company successfully bidded for the three properties and discovered the true situation.

to facts present at the time of contract, while frustration applies to supervening events. This creates some confusion, as the courts have tended to treat the encountering of unforeseen ground conditions as a supervening event, although the conditions will normally have existed at the time of contract.

The heading mistake refers to situations where one or both parties to a contract are under a misapprehension of present fact at the time of contract. The situations can be divided for the purpose of legal analysis into two categories namely; unilateral mistake, and common mistake.

Unilateral mistake where the mistake is such that the parties are at cross-purpose, or where the mistaken belief of one party is known to the other party. The doctrines relating to unilateral mistake are essentially extensions of the doctrines on formation of contract and the need for agreement. Because of the lack of true agreement, the contract will be rescinded and it is not difficult to say that the contract never really existed. There is a distinction between the various situations of unilateral mistake, as to the relevant test applicable. Where the mistake is known to one party, the test is subjective as to the belief of the mistaken person. When the parties are at cross purposes, the test is objective and relates to what a reasonable person would have understood from the agreement as expressed.

Common mistake, also called mutual mistake, where the parties share the same misapprehension. This includes cases where one party knows of the mistake but is unaware of, or does not consider, its significance. There is some disagreement about this terminology. Where the parties share a common misapprehension, the position is different as there is clearly agreement. The law of common mistake sets out two conditions that are required for a contract to be held inoperative; firstly, the mistake should be sufficiently fundamental or basic to invalidate a contract, and secondly, the mistake must not be the responsibility of one or other of the parties.

Law cases show relief, granted in the case of a mistake in law, in four main situations:

1. In all situations the mistake must be one of fact.
2. Where there has been a mistake as to the identity of the other party contracting and the first party did not intend to enter, and would not have entered, into a contract with that person.
3. A party signing a contract document has been misled as to the nature of the document, in which case the plea of '*non est factum*' may apply.
4. Where there is clear mistake by the offerer as to the terms of the contract, and the mistake is known to the other party when he purports to accept.

Tamplin v James[19] is a contract law case concerning the availability of specific performance for a breach of contract induced by mistake. The case established that if a person enters a contract on the basis of a mistake that was not induced by the other party to the contract, specific performance will be awarded against the person

---

[19] (1880) 15Ch D 215.

if no hardship amounting to injustice would be inflicted on the person by holding the person to the contract. LJ Baggalay observed that: "Where there has been no misrepresentation and where there is no ambiguity in the terms of the contract, the defendant cannot be allowed to evade the performance of it by the simple statement that he has made a mistake. Were such to be the law the performance of a contract could seldom be enforced upon an unwilling party who was also unscrupulous."

In *Bell v Lever Bros*,[20] Lord Atkin's judgment is generally regarded as one of the leading pronouncements on the law of mistake as he said:

> Whenever it is to be inferred from the terms of a contract or its surrounding circumstances that the consensus has been reached upon the basis of a particular contractual assumption, and that assumption is not true, the contract is avoided: i.e. it is void ab initio, and it ceases to bind if the assumption is of future fact. A mistake will not affect assent unless it is the mistake of both parties and is as to the existence of some quality which makes the thing without the quality essentially different from the thing it was believed to be.

In *Bell v Lever Bros,* Lord Thankerton defined common mistake as:

> [It] can only properly relate to something which both must necessarily have accepted in their minds as an essential and integral part of the subject-matter. Logically, before one can turn to the doctrines as to mistake, whether at common law or in equity, one must first determine whether the contract itself, by express or implied condition precedent or otherwise, provides who bears the risk of the relevant mistake. It is at this hurdle that many pleas of mistake will fail or prove to have been unnecessary. Only if the contract is silent on the point is there scope for invoking mistake.

In *McRae v Commonwealth Disposals Commission*,[21] where the Commission had invited tenders for the purchase of a sunken tanker which had never, in fact, existed; it was held that the successful tenderer was entitled to recover his abortive costs on the basis of an implied warranty as to the existence of the tanker. J. Fullagar said:

> Whether the contract is void for common mistake is primarily a matter of construction. Was there an implied condition precedent that the goods were in existence? In this situation, one can only conclude that the Commonwealth promised that there was a tanker in the position specified. Whether the court will imply a warranty in respect of the matter concerning which both parties were mistaken, will depend on whether the warrantor had expertise and special information available. In situations where both parties were equally able to check the situation out, then the courts will be less reluctant to find that matter warranted.

---

[20] Bell v Lever Brothers Ltd (1932) AC 161. Lever had terminated the employment of Bell, who was chairman of their operation in Nigeria, and had agreed to make him a payment as compensation. Bell had agreed the payment and payment had been made to him. Lever then discovered that he had been involved in corrupt activities which would have entitled them to dismiss him summarily without payment, and they brought an action for rescission of the agreement and to recover the payment.

[21] (1951) 84 CLR 77.

In *Sindall v Cambridgeshire CC*,[22] the court concluded on the facts of the case that the essential obligation was not impossible by reason of the mistake. LJ Hoffman, in his *dicta*, held there were no misrepresentation and no operative mistake. He said that the three factors for deciding what is equitable in the case of mistake are as follows:

1. The nature of the mistake. This means that the court was meant to consider the importance of the representation in relation to the subject matter of the transaction.
2. The loss that would be caused by the mistake if the contract were upheld.
3. The loss that rescission would cause to the other party.

In *Sindall v Cambridgeshire CC*, J. Steyn, in regard to the principle of mistake, further said:

> In my judgment a party cannot be allowed to rely on a common mistake where the mistake consists of a belief which is entertained by him without any reasonable grounds for belief. That is not because principles such as estoppel or negligence dictate it, but simply because policy and good sense dictate that the positive doctrines regarding common mistake should be so qualified.

In *Great Peace Shipping v Tsavliris Salvage*,[23] the Court of Appeal reviewed and explained *Bell v Lever Bros* (1932) (*supra*), and stressed the linkage between frustration and mistake. Lord Phillips stated:

> At the time of Bell v Lever Bros the law of frustration and common mistake had advanced hand-in-hand on the foundation of a common principle. Thereafter frustration proved a more fertile ground for the development of this principle than common mistake, and consideration of the development of the law of frustration assists with the analysis of the law of common mistake. The avoidance of a contract on the ground of common mistake results from a rule of law under which, if it transpires that one or both parties have agreed to do something which it is impossible to perform, no obligation arises out of that agreement.

---

[22] William Sindall Plc v Cambridgeshire County Council (1994) 3 All ER 932. William Sindall agreed to buy land from Cambridgeshire County Council after they were told the council was aware of no easements. However, a private sewer from 20 years before was found after completion. In that case, the claim was to set aside a contract for the sale of land on the grounds of mistake. The existence of a sewer succeeded at first instance, but was rejected by the Court of Appeal.

[23] Great Peace Shipping Ltd. v Tsavliris Salvage (International) Ltd. (The Great Peace) (2002) 4 All ER 689. The defendant agreed to provide salvage services for a stricken vessel. The brokers were informed that a vessel, owned by the claimant should be able to reach the stricken vessel within about 12 h. In fact, unbeknown to either party, the two vessels were some 410 miles apart, and it would have taken the claimant's vessel 39 h to reach the stricken vessel.

3.7 Frustration

Under law, frustration sometimes provides relief where the performance or purpose of the contract has been made impossible or illegal by supervening events (i.e. events taking place after the contract was performed), which were not foreseen at the time of contract and which were not due to any act or omission of either party. The consequence of the doctrine is that, in the event of frustration, the contract will be discharged in relation to the future performance obligations of both parties. A claimant must firstly establish whether or not the particular situation in question has been expressly provided for in the contract under a *force majeure* clause. For example, a construction contract might include specific provisions for weather conditions or acts of civil disobedience. A *force majeure* clause is only valid if the provision is full and complete—that is, it has to be specific about what risk is being provided for.

Frustration of a contract occurs only where after the conclusion of the contract a fundamentally different situation has unexpectedly emerged. The emergence of some new set of circumstances may make the performance of the contract more difficult, onerous or costly than what was envisaged by the parties when entering into the contract; for example, a sudden, even abnormal, rise or fall in material prices or the failure of a particular source of supply requiring the contractor to obtain supplies from another more expensive source. However, these events will not normally operate to frustrate a contract unless they are of a proportion to make the performance of the contract impossible.[24]

Occurrence of the frustrating event brings the contract to an end forthwith, without more and automatically. The establishment of a satisfactory mechanism for the courts to provide relief in extreme cases, assuming the need for such a mechanism is accepted, is made difficult by the principle of law that the courts will not rewrite an existing contract. In order to allow adjustment, it is necessary first to kill off the original contract. This creates a rather blunt instrument for making adjustment.

There is, effectively, a remedy in some cases of frustration in construction contracts that if the original contract is terminated by frustration, but the ultimate object can still be achieved by some other means and the parties agree so to proceed, then, in the absence of an agreement on price, a *quantum meruit* will be payable in respect of work performed after the original contract was terminated. This remedy has been invoked and sometimes allowed by the courts as a means to provide relief to construction contractors encountering unforeseen difficulties.

The doctrines of frustration and the impossibility of performance were first recognised as an excuse for a party's failure to perform in the late nineteenth century. Traditionally, courts have applied this doctrine narrowly due to a judicial

---

[24] Unlike English law, the United States law has abandoned the word 'impossible' and used the term 'impracticable'. The impracticability of performance of contract includes situations of extra and unreasonable difficulty, expenses, injury or loss to one of the parties.

recognition that the purpose of contract law is to allocate risks and that failure to perform should only be excused in extreme cases. In the words of one unreported court case, "impossibility excuses a party's performance only when the destruction of the subject matter of the contract or the means of performance makes performance objectively impossible. Moreover, the impossibility must be produced by an unanticipated event that could not have been foreseen or guarded against in the contract." Most courts, since, have recognised that the defence of frustration is too harsh in its requirement that performance be absolutely impossible and have instead moved towards a standard of commercial impracticability.

In *Taylor v Caldwell*,[25] the doctrine of frustration was extended to perishing of things essential to performance. In that case, a theatre, hired for a series of concerts and fetes, was burnt down. It was held that the contract was discharged. J. Blackburn identified the basis of frustration more generally as follows:

> Where there is a positive contract to do a thing not in itself unlawful, the contractor must perform it or pay damages for not doing it, although in consequence of unforeseen accidents, the performance of his contract has become unexpectedly burthensome or even impossible. But this rule is only applicable where the contract is positive and absolute, and not subject to any condition either express or implied. Where from the nature of the contract, it appears that the parties must from the beginning have known it could not be fulfilled unless when the time for fulfillment of the contract arrived some particular specified thing continued to exist, so that, when entering into the contract, they must have contemplated such continuing existence as the foundation of what was to be done; there in the absence of any express or implied warranty that the thing shall exist, the contract is not to be construed as a positive contract, but subject to an implied condition that the parties shall be excused in case, before breach, performance becomes impossible from the perishing of the thing without the default of contractor.

In *Davis Contractors v Fareham UDC*,[26] the contractors argued that the contract was frustrated and they could therefore claim on a *quantum merit* basis, which would be more than the contract price, for the houses they completed. Lord Radcliffe formulated the classic statement of the modern doctrine of frustration as follows:

> The theory of frustration belongs to the law of contract and it is represented by a rule which the courts will apply in certain limited circumstances for the purpose of deciding that contractual obligations, ex facie binding, are no longer binding on the parties. Frustration occurs whenever the law recognises that without default of either party a contractual obligation has become incapable of being performed because the circumstances in which performance is called for would render it a thing radically different from that which was undertaken by the contract. Non haec in foedera veni. It was not this that I promised to do. It is not hardship or inconvenience or material loss itself which calls the principle of frustration into play. There must be as well such a change in the significance

---

[25] (1863) 3 B & S 826.

[26] Davis Contractors Ltd. v Fareham Urban District Council (1956) AC 696. On July 9 1946. The contractors entered into a building contract to build 78 houses for a local authority for £92,425 within a period of eight months. Without the fault of either party, adequate supplies of labour were not available and the work took 22 months to complete.

of the obligation that the thing undertaken would, if performed, be a different thing from that contracted for.

Where a contract is frustrated, the parties are discharged from future performance, but accrued rights and liabilities stand. At common law, this could work harshly either way as it depended whether or not an entitlement of a payment has accrued. Thus, by either relieving a party completely of its future performance obligations or effectively enabling a re-pricing of the work, frustration is sometimes used as a means for providing relief from contractual obligations which have become exceptionally onerous. Frustration, as Lord Roskill stated in '*The Nema*',[27] is "not lightly to be invoked to relieve contracting parties of the consequence of imprudent commercial bargains." Further, in '*The Nema*', Lord Diplock, in the identification of a frustrating event, stated the following: "Never a pure question on fact but does in the ultimate analysis involve a conclusion of law as to whether the frustrating event or series of events has made the performance of the contract a thing radically different from that which was undertaken by the contract."

In *J Lauritzen AS v Wijsmuller BV*,[28] the Court of Appeal held that the contract could not be cancelled under clause 17[29] and that the doctrine of frustration did not operate on the facts of the case because the defendants could have fulfilled their contractual obligations by using *Super Servant One*. Additionally, the defendants would be precluded from relying on the doctrine of frustration if it could be shown that the loss of *Super Servant Two* had been caused by their own negligence. LJ Bingham, to relieve a party from its obligations to perform under the doctrine of frustration, proposed the following conditions:

> First, the doctrine of frustration was evolved to mitigate the rigour of the common law's insistence on literal performance of absolute promises. The object of the doctrine was to give effect to the demands of justice, to achieve a just and reasonable result, to do what

---

[27] Pioneer Shipping Ltd. v BTP Tioxide Ltd. (The Nema) (1982) AC 724. After one round voyage the Nema arrived back at Sorel on June 20, 1979. She gave notice of readiness but was unable to load owing to a strike.

[28] J Lauritzen AS v Wijsmuller BV (The Super Servant Two) (1990) 1 Lloyds Rep 1. The defendants, Wijsmuller, agreed to carry Lauritzen's drilling rig (the Dan King) from a shipyard in Japan to a delivery location off the coast of the Netherlands. The rig was to be delivered using either the Super Servant One or the Super Servant Two. Super Servant Two sank. The defendants informed the plaintiffs that they could no longer carry out the Dan King contract, claiming that they were permitted to cancel the contract under clause 17 and that, in any event, the contract had been frustrated by the sinking of Super Servant Two.

[29] Clause 17: 'Wijsmuller has the right to cancel its performance under this Contract whether the loading has been completed or not, in the event of force majeure(sic), Acts of God, perils or danger and accidents of the sea, acts of war, warlike-operations, acts of public enemies, restraint of princes, rulers or people or seizure under legal process, quarantine restrictions, civil commotions, blockade, strikes, lockout, closure of the Suez or Panama Canal, congestion of harbours or any other circumstances whatsoever, causing extra-ordinary periods of delay and similar events and/or circumstances, abnormal increases in prices and wages, scarcity of fuel and similar events, which reasonably may impede, prevent or delay the performance of this contract.'

was reasonable and fair, as an expedient to escape from injustice where such would result from enforcement of a contract in its literal terms after a significant change in circumstances. Secondly, since the effect of frustration is to kill the contract and discharge the parties from further liability under it, the doctrine is not to be lightly invoked, but must be kept within very narrow limits and ought not to be extended. Thirdly, frustration brings the contract to an end forthwith, without more and automatically. Fourthly, the essence of frustration is that it should not be due to the act or election of the party seeking to rely on it. A frustrating event must be some outside event or extraneous change. Fifthly, a frustrating event must take place without blame or fault on the side of the party seeking to rely on it.

In *McAlpine Humberoak v McDermott International*,[30] the Court of Appeal rejected frustration as a means to relief. The court held that, faced with a significant number of change orders, McAlpine's original lump sum contract had become frustrated, giving rise to a substituted contract under which the price to be paid should be calculated on *quantum meruit* based on *'cost plus'* basis. LJ Lloyd commented: "If we were to uphold the judge's finding of frustration, this would be the first contract to have been frustrated by reason of matters which had not only occurred before the contract was signed, and were not only well known to the parties but had also been provided for in the contract itself."

3.8 Economic Duress

Duress is a means by which a person may be released from the obligations under a contract where unlawful threats have been made. The duress is of the sort that deprives the person of consent in entering the contractual arrangement, although the test at law is that the person is exposed to pressure and is deprived of choice that would otherwise be available. There must be effectively no choice other than to comply with the request or the demand to be successful in a claim for duress. In the commercial context this vitiating factor may be alleged where illegitimate pressure has been made that would affect a person's economic interests. A typical example in construction is where the employer withholds payments for the contractor unless certain tasks are completed.

In construction, a contract entered into under economic duress is voidable and not void. A contractor or the client who has entered into the contract may either affirm or avoid such contract after the duress has ceased; and if he has so voluntarily acted under it with the full knowledge of all the circumstances he may be held bound on the ground of ratification, or if, after escaping from this vitiating factor, he takes no steps to set aside the formed agreement he may be found to have affirmed it.

Economic duress may apply to the formation of the contract, at the commencement of the performance of the contract or subsequent variations of the

---

[30] (1992) 58 BLR 1.

contract. The pressure must be of a nature that is illegitimate and was a significant cause inducing the person to agree to the terms of the contract. The threats made and pressure asserted must be particularly coercive and of some significant weight or gravitas. The injured party conduct must be affected in a significant way by the duress, and a reasonable alternative must not be available at the time of the duress.

Economic duress is characterised by a lack of choice. Where an alternative is available to the injured party, the vitiating defence will not be available, however, the alternative must be reasonable. The following descriptive examples may give rise to a claim for economic duress:

1. Threats to terminate a contract, where the threat is properly regarded as illegitimate pressure.
2. Applying pressure in bad faith.
3. Making threats that are calculated to seriously damage another.
4. Threats to prosecute where the charge is known to be false.
5. Requirements for extra payments to be made over and above the original contract price.
6. Using knowledge of the affairs of the person suffering the duress to apply illegitimate pressure.

In *North Ocean Shipping v Hyundai*,[31] the court held, *obiter dicta*, that the recovery of money on the ground that it had been paid under duress, other than under duress to the person, was not limited to cases where there had been duress to goods; the duress could also take the form of economic duress, which could be constituted by a threat to break a contract. If, however, a party who had entered into a contract under economic duress later affirmed the contract, he was then bound by it. In this case, J. Mocatta opined that conduct does not have to be tortuous to constitute duress for the purpose of law; however, to amount to duress there must be more than mere commercial pressure. He stated the following: "A threat to break a contract may amount to such 'economic duress'. If there has been such a form of duress leading to a contract for consideration, I think that contract is a voidable one which can be avoided and the excess money paid under it recovered."

In the case of *Pao On v Lau Yiu Long*,[32] the court held that the issue at stake was commercial pressure and no more, since, in reality, the company just wanted to avoid adverse publicity. Lord Scarman held that coercion of will depends upon

---

[31] North Ocean Shipping Co. Ltd v Hyundai Construction Co Ltd (1979) QB 705. The defendants agreed to build a tanker for the plaintiffs at a price fixed in U.S. dollars to be paid by instalments. After the first instalment the U.S. dollar was devalued by 10 per cent and the defendants insisted on further instalments being increased by 10 per cent. The plaintiffs refused and suggested arbitration, but it became apparent that the defendants would not continue their work without this agreement.

[32] (1980) AC 614. This case relates an indemnity that Pao On would be indemnified if the shares were worth less than $2.50 each. The share price fell below, and the Lau's refused to indemnify.

the individual circumstances in each case; however, he suggested that the following factors may also be considered in asserting economic duress by stating:

> Duress, whatever form it takes, is a coercion of the will so as to vitiate consent. In a contractual situation commercial pressure is not enough. In determining whether there was a coercion of will such that there was no true consent, it is material to enquire whether the person alleged to have been coerced did or did not protest; whether, at the time he was allegedly coerced into making the contract, he did or did not have an alternative course open to him such as an adequate legal remedy; whether he was independently advised; and whether after entering the contract he took steps to avoid it.

In *Dimskal Shipping v ITF*,[33] the court considered the developing law of economic duress and held, *obiter dicta*, that the question of whether economic pressure constituted duress of such a kind as to entitle the innocent party to avoid the contract is to be determined by reference to the proper law of the contract. In order to justify avoidance of a contract, the economic pressure must be such as to be called illegitimate. Lord Goff, in affirming Lord Scarman test in *Pao On v Lau Yiu Long* (1980) (*supra*), added another factor in asserting the doctrine of economic duress as: "It must be shown that the payment made or the contract entered was not a voluntary act."

3.9 Foreseeability

The exploration of concepts such as foreseeability in the common law, even in different contexts, can illuminate philosophical issues which arise in construction. Although the common law refuses to limit the extent of performance obligations based on what was foreseeable at the time of contract, the concept of foreseeability is found in several common law contexts; first, it appears in relation to remoteness of damages in contract and tort and second, in the context of the existence or extent of a duty of care in the tort of negligence. The basic normative problem is to determine what level of knowledge the acting person shall have when he makes a list of all possible consequences of his intended action and to determine what degree of probability he should require in order to say that the damage was a computable function of the negligent act.

Any person who suffers a loss caused by another person's negligence shall be compensated. However, compensation is granted only to the extent that the loss was a foreseeable consequence of the negligent act (or omission). When a detrimental effect is said to be foreseeable, in this legal sense, it means that the effect is a computable function of an action. However, many courts impose an outright

---

[33] Dimskal Shipping Co SA v International Transport Workers Federation (The Evia Luck) (1991) 4 All ER 871. The plaintiff shipowners had been induced by industrial action against a vessel in Sweden. One of the documents signed provided that the undertaking was to be governed by English law. The plaintiffs purported to avoid the agreements for duress and to recover the monies that they had paid under them.

requirement that the event in question was unforeseeable at the time the parties entered into the contract. Where unforeseeability is a requirement, the relevant inquiry is whether the contingency in question was so unusual or unforeseen, and the consequences so severe, that to require performance by the promisor would grant the promise an advantage he did not bargain for.

The rule of remoteness in contract, derived from *Hadley v Baxendale*,[34] is that damages for breach of contract are only recoverable insofar as they are of a type that was foreseeable at the time of contract. This case, therefore, raises sharply the question as 'to the nature and extent of the duty of a client whose contract operations may cause losses to contractors executing the works?' The answer to this question, depending on the facts of the case, is twofold; it might be one of who he must not carry out or permit an operation which he knows or ought to know clearly can cause such losses, however improbable that result may be, or that he is only bound to take into account the possibility of such damage if such damage is such that a reasonable client careful of the operations of the project being executed, would regard that risk as material.

In *British Movietonews v London & District Cinemas*,[35] the court held, *obiter dicta*, that there is, however, no general principle at common law that performance obligations are restricted by foreseeability; a party is not relieved from completing an obligation, nor is he entitled to additional payment, merely because the obligation proves more difficult or onerous than what was foreseen or foreseeable. Viscount Simon, in identifying when a contract ceases to operate when facing unprecedented conditions, said:

> If, on the other hand, a consideration of the terms of the contract, in the light of the circumstances existing when it was made, shows that they never agreed to be bound in a fundamentally different situation which has now unexpectedly emerged, the contract ceases to bind at that point-not because the court in its discretion thinks it just and reasonable to qualify the terms of the contract, but because on its true construction it does not apply in that situation.

In *British Movietonews v London & District Cinemas,* LJ Denning, by holding that the courts will not release a party from a bargain merely because it does not turn out as anticipated, observed:

> No matter that a contract is framed in words which taken literally or absolutely, cover what has happened, nevertheless, if the ensuing turn of events was so completely outside the contemplation of the parties that the court is satisfied that the parties, as reasonable people, cannot have intended that the contract should apply to the new situation, then the court will read the words of the contract in a qualified sense; it will restrict them to the circumstances contemplated by the parties; it will not apply them to the uncontemplated turn of events, but will do therein what is just and reasonable.

---

[34] (1854) 9 Exch 341.

[35] (1952) AC 166.

In *Bende & Sons, Inc. v Crown Recreation, Inc.*,[36] the reasoning the court employed was essentially an assumption of risk argument, and that the foreseeability element is to probe the '*basic assumption*' requirement. J. Mclaughlin stated that the defendant must demonstrate either that "the contingency that made performance impracticable was not foreseeable at the time of contracting or the contract contains specific, exculpatory language excusing non-performance under certain circumstances." He further stated: "The foreseeability requirement does not entail contemplation of a specific contingency; rather, it is sufficient that the contingency that eventually occurred could have been foreseen as a real possibility that would affect performance."

3.10 Performance and Breach of Obligations

The terms of a contract amount to promises by each party. Some will be promises as to existing fact but mostly, they will be promises to be performed. Typically, the promise to be performed may consist of making a payment, supplying goods or carrying out work. The general principle is sometimes expressed by the maxim *pacta sunt servanda* which stipulates that agreements are to be observed or kept. Most legal textbooks state that, "performance of a promise must be precise and exact" and that "any positive obligation, whatever its source, is extinguished by being performed." It is also considered that rights to payment may be dependent on performance. In particular some contracts are considered to be entire so that no entitlement to payment accrues until the entire obligation has been performed.

A party in a contract is entitled to damages that will permit him to complete that which he contracted for as he intended it to be completed. However, where the cost of completion is grossly and unfairly disproportionate to the good to be attained, the measure of damages is the difference in value. Generally, the law establishes that if a party's contractual performance has failed to provide to the other contracting party something to which that other was, under the contract, entitled, and which, if provided, would have been of value to that party, then, if there is no other way of compensating the injured party, the injured party should be compensated in damages to the extent of that value.

---

[36] 548 F. Supp. 1018, 1022 (E.D.N.Y. 1982). Bende contends that it had a contract to supply the Government of Ghana with 10,000 pairs of boots; the Kiffe, which is a division of the defendant Crown Recreation, Inc., agreed to manufacture the boots in Korea and to deliver them in Ghana; the Kiffe failed to deliver the boots on the agreed date; and that as a result of this failure Bende suffered $44,685 in damages when the Government of Ghana cancelled its resale contract.

In *Jacob & Youngs v Kent*,[37] the court held, *obiter dicta*, that equity and fairness dictate that one who unintentionally commits a trivial wrong will not be condemned to a fate so clearly out of proportion with the transgression. To permit the defendant to recover the cost of replacement of the pipe would be unduly oppressive. Instead, the defendant will be adequately compensated by recovering the difference in value of a home with the specified pipe and the value of the home, as it exists, with a different kind of pipe. In asserting this *dictum*, J. Cardozo said:

> We think the evidence, if admitted, would have supplied some basis for the inference that the defect was insignificant in its relation to the project. The courts never say that one who makes a contract fills the measure of his duty by less than full performance. They do say, however, that an omission, both trivial and innocent, will sometimes be atoned for by allowance of the resulting damage, and will not always be the breach of a condition to be followed by a forfeiture. In the circumstances of this case, we think the measure of the allowance is not the cost of replacement, which would be great, but the difference in value, which would be either nominal or nothing.

In summary, in the case of *Jacob & Youngs v Kent*, the court affirmed the following criteria in terms of performance and damages due in case of breach:

- The contractor default was unintentional and trivial, and that they had substantially performed on the contract.
- The client was entitled to recover the difference in the value of the house resulting from the use of a different brand of pipe (if any), but other than that, he was required to pay the full amount of the contract.
- The breach was not a condition of the contract. Breaches that are not conditions are atoned for by calculating damages, they do not excuse the other party from performance.

In *Radford v De Froberville*,[38] the plaintiff built a dividing wall on his own land, which the defendant was supposed to build through an agreement between the two, and claimed the cost of doing so from the defendant. The defendant maintained that the appropriate measure of damages was the consequent diminution in the value of the plaintiff's property, which was nil. In this case, J. Oliver said:

---

[37] (1921) 121 NE 889. The plaintiff built a house for defendant for a price of $77,000, and sued to recover the balance due of $3,483.46. Defendant specified that all pipes in the house must be Reading pipe, but inadvertently, plaintiff installed pipes that was not Reading pipes. When defendant discovered this defect, he demanded that the work be redone, which would have required the demolition and reconstruction of substantial parts of the house. Kent refused to pay and Jacob & Youngs initiated this action.

[38] (1977) 1 WLR 1262. A contract was made for the sale of a plot of land adjoining a house belonging to the plaintiff (the vendor) but occupied by his tenants, under which the defendant (the purchaser) undertook to build a house on the plot and also to erect a wall to a certain specification on the plot so as to separate it from the plaintiff's land. The plaintiff obtained judgment against the defendant for damages for breach of contract by reason of failure to erect the dividing wall, but an issue arose as to the measure of the damages.

If he contracts for the supply of that which he thinks serves his interests, be they commercial, aesthetic or merely eccentric, then if that which is contracted for is not supplied by the other contracting party I do not see why, in principle, he should not be compensated by being provided with the cost of supplying it through someone else or in a different way, subject to the proviso, of course, that he is seeking compensation for a genuine loss and not merely using a technical breach to secure an uncovenanted profit.

The principles of performance, breach of obligations and the damages that are due came under pressure in *Ruxley Electronics v Forsyth*.[39] The plaintiff claimed, as damages, the cost of reinstating the pool to the depth stipulated in the contract. The court contemplated the fact that to award the full cost of replacement as damages was manifestly unreasonable, but to say that the plaintiff had to make do with what had been supplied to him, in disregard of the specification, was unacceptable.

In *Ruxley Electronics v Forsyth*, a majority in the Court of Appeal held that the defendant was entitled to the full cost of replacement as damages, based on his entitlement to have the contract performed as stipulated. The court, then, held that the intention of defendant as to what he would actually do with the damages was said to be of no concern to the plaintiff. The House of Lords, however, held that the recovery of damages was subject to a requirement of reasonableness, that Mr Forsyth was not entitled to the cost of replacement, but he was entitled to a figure of £2,500 which had been awarded by the original trial judge for "loss of amenity". Lord Bridge said:

Damages for breach of contract must reflect, as accurately as the circumstances allow, the loss which the claimant has sustained because he did not get what he bargained for. There is no question of punishing the contract breaker. Given this basic principle, the court, in assessing the measure of the claimant's loss has ultimately to determine a question of fact, although the law has of course developed detailed criteria which are to be applied in ascertaining the appropriate measure of loss in a wide variety of commonly occurring situations. Since the law relating to damages for breach of contract has developed almost exclusively in a commercial context, these criteria normally proceed on the assumption that each contracting party's interest in the bargain was purely commercial and that the loss resulting from a breach of contract is measurable in purely economic terms. But this assumption may not always be appropriate. The circumstances giving rise to the present appeal exemplify a situation which one might suppose to be of not infrequent occurrence. A landowner contracts for building works to be executed on his land. When the work is complete it serves the practical purpose for which it was required perfectly satisfactorily. But in some minor respect the finished work falls short of the contract specification. The difference in commercial value between the work as built and the work as specified is nil. But the owner can honestly say: 'This work does not please me as well as would that for which I expressly stipulated. It does not satisfy my personal preference. In terms of amenity, convenience or aesthetic satisfaction I have lost something.' Nevertheless the

---

[39] (1996) AC 344. The plaintiff builders had entered into a contract with the defendant, Mr Forsyth, to construct a swimming pool in the grounds of his house for the sum of £17,800. The tender had originally been for a pool with a depth of 6ft 9ins at the deep end, but Mr Forsyth had negotiated with the builders that the depth would be increased to 7ft 6ins at no extra cost. The pool was built, but disagreement arose. Mr Forsyth refused to pay the balance of the price, so the builders sued.

contractual defect could only be remedied by demolishing the work and starting again from scratch. The cost of doing this would be so great in proportion to any benefit it would confer on the owner that no reasonable owner would think of incurring it.

In *Ruxley Electronics v Forsyth*, Lord Mustill referred to situations where, in carrying out of building works, there had been minor deviations from the contractual specifications but where the deviations had not reduced the value of the property below the value it would have had if the work had been properly carried out. He stated:

> Yet the householder must surely be entitled to say that he chose to obtain from the builder a promise to produce a particular result because he wanted to make his house more comfortable, more convenient and more conformable to his own particular tastes; not because he had in mind that the work might increase the amount which he would receive if, contrary to expectation, he thought it expedient in the future to exchange his home for cash. To say that in order to escape unscathed the builder has only to show that to the mind of the average onlooker, or the average potential buyer, the results which he has produced seem just as good as those which he had promised would make a part of the promise illusory, and unbalance the bargain.

3.11 Contract Termination

Rescission in its most basic form is where neither party has performed the whole of his obligations and there is a mutual agreement to rescind the contract. Such a rescission can be expressed or implied. In the context of breach of contract, a party is said to rescind a contract when he lawfully terminates both parties' future performance obligations under the contract. Rights and liabilities which have already accrued at the time of rescission are not affected.

The word rescission means, in the context of misrepresentation and mistake, that the contract is effectively unravelled so as to return to the situation as if it had never existed. The purpose of rescission is to put the parties in the position as if the contract had never been entered, i.e., the contract is said to be rescinded *ab initio*.

A misrepresentation, even one that was incorporated into the contract, gives the innocent party the option of rescinding the contract. The misrepresentation must be material, substantial or go to the root of the contract.

The party asserting mistake as grounds to rescind a contract must show that both parties were mistaken as to a material matter at the time the contract was entered into. Courts require the plaintiff to prove the following elements for the court to grant rescission based on mutual mistake:

1. The mistake must have existed at the time the contract was entered into.
2. The mistake must have been mutual and common to all the parties.
3. The mistake must have involved a material matter.
4. The mistake must have been such that the parties intended to say one thing but by the written instrument expressed another.

Problems have arisen from the misuse of the word rescission to describe an accepted repudiation. To use these terms synonymously can only lead to confusion and should be avoided. Rescission is a remedy available to the representee, *inter alia*, when the other party has made a false or misleading representation or in case of mistake. Repudiation, by contrast, occurs by words or conduct evincing an intention not to be bound by the contract. Contrary to rescission, which allows the rescinding party to treat the contract as if it were void *ab initio*, the effect of repudiation depends on the election made by the non-repudiating party. If the non-repudiating party accepts the repudiation, the contract is terminated, and the parties are discharged from future obligations, although rights and obligations that have already matured are not extinguished. If the repudiation is not accepted, the contract remains in being for the future and each party has the right to sue for damages for past or future breaches. The word renunciation is also used to mean repudiation.

Thus, it is said in *Chitty on Contracts*:

> The question whether a rescission has been effected is frequently one of considerable difficulty, for it is necessary to distinguish a rescission of the contract from a variation which merely qualifies the existing rights and obligations. If a rescission is effected the contract is extinguished; if only a variation, it continues to exist in an altered form. The decision on this point will depend on the intention of the parties to be gathered from an examination of the terms of the subsequent agreement and from all the surrounding circumstances. Rescission will be presumed when the parties enter into a new agreement which is entirely inconsistent with the old, or, if not entirely inconsistent with it, inconsistent with it to an extent that goes to the very root of it. The change must be fundamental and the "question is whether the common intention of the parties was to 'abrogate', 'rescind', 'supersede', or 'extinguish' the old contract by a 'substitution' of a 'completely new' or 'self-subsisting' agreement". A renunciation of a contract occurs when one party by words or conduct evinces an intention not to perform, or expressly declares that he is or will be unable to perform, his obligations under the contract in some essential respect.

A useful definition of rescission comes from Lord Atkinson in *Abram Steamship Co. v Westville Shipping Co.*[40]:

> Where one party to a contract expresses by word or act in an unequivocal manner that by reason of fraud or essential error of a material kind inducing him to enter into the contract he has resolved to rescind it, and refuses to be bound by it, the expression of his election, if justified by the facts, terminates the contract, puts the parties in status quo ante and restores things, as between them, to the position in which they stood before the contract was entered into.

The consequences when a contract is brought to an end by the acceptance by one party to it of a repudiatory breach of contract by the other party are well established. They were clearly stated by J. Dixon in *McDonald v Dennys Lascelles Ltd.*,[41] where he said:

---

[40] (1923) AC 773.
[41] (1933) 48 CLR 457.

When a party to a simple contract, upon a breach by the other contracting party of a condition of the contract, elects to treat the contract as no longer binding upon him, the contract is not rescinded as from the beginning. Both parties are discharged from the further performance of the contract, but rights are not divested or discharged which have already been unconditionally acquired. Rights and obligations which arise from the partial execution of the contract and causes of action which have accrued from its breach alike continue unaffected.

In *Keneric Tractor Sales Ltd. v Langille*,[42] J. Wilson addressed the distinction between rescission and repudiation as follows:

The modern view is that when one party repudiates the contract and the other party accepts the repudiation the contract is at this point terminated or brought to an end. The contract is not, however, rescinded in the true legal sense, i.e. in the sense of being voided ab initio by some vitiating element. The parties are discharged of their prospective obligations under the contract as from the date of termination but the prospective obligations embodied in the contract are relevant to the assessment of damages.

In *Woodar Developments v Wimpey*,[43] the question arose whether an unjustified rescission by Wimpey amounted to repudiation. It was held by the House of Lords that Wimpey were actually relying on the contract (although erroneously) rather than refusing to be bound by it, and since their conduct showed that they intended to abide by the court's interpretation of the contract if it went against them, their unsuccessful attempt at rescission did not amount to repudiation. In reaching his decision, Lord Wilberforce emphasised that, in considering whether there has been a repudiation, it is necessary to consider all the circumstances and in particular, the contract breaker's conduct as a whole, and whether it indicates an intention to abandon and to refuse performance of the contract. He further commented that "unless the invocation of that provision were totally abusive, or lacking in good faith, the fact that it has proved to be wrong in law cannot turn it into a repudiation."

As regards to future primary obligations, the general principle was stated by Lord Diplock in *Photo Productions v Securicor*[44] that where a contract is treated as discharged, the effect is that primary obligations as to future performance are terminated. A secondary obligation is substituted in their place, by implication of law, that the party in default should pay monetary compensation to the other party for the loss sustained by him in consequence of the non-performance of the future obligations. Generally speaking, a valid termination of the contract releases not

---

[42] (1987) 2 SCR 440.

[43] (1980) 1 All ER 571. In *Woodar v Wimpey*, in which the House of Lords found a rescission to be wrongful but held there had been no repudiatory breach because the conduct of the contract breaker did not of itself manifest an intent to breach the contract, instead termination was purportedly effected under the agreement itself.

[44] Photo Production Ltd v Securicor Transport Ltd (1980) AC 827. Securicor Transport agreed to provide a night patrol service for Photo Production's factory to protect from theft and fire etc. An employee of Securicor Transport, while supposed to be patrolling the premises, lit a fire (to keep warm) and ended up burning the factory down.

only the victim of breach but also the party in breach from their primary obligations to perform in the future. However, the defaulting party is not totally discharged from any liabilities, but may be liable to pay damages and that liability may relate both to breaches committed before termination and to losses suffered by the injured party as a result of the defaulting party's repudiation of future obligations.

In *Eminence Property Developments Ltd. v Heaney*,[45] the court clarified the test for repudiatory breach of contract. In this case, the court found that a vendor of land had not acted in repudiatory breach of contract where, by mistake, he served notices of rescission on the purchaser before the final date for complying with notices to complete had been reached. In the Court of Appeal, LJ Etherton confirmed the legal test in respect of repudiatory conduct as such:

1. Whether looking at the circumstances objectively that is, from the perspective of a reasonable person in the position of the innocent party, the contract breaker has clearly shown an intention to abandon and altogether refuse to perform the contract.
2. Whether or not there has been a repudiatory breach is highly fact sensitive and that is why comparison with other cases is of limited value.
3. All the circumstances must be taken into account insofar as they bear on an objective assessment of the intention of the contract breaker. This means that motive, while irrelevant if relied upon solely to show the subjective intention of the contract breaker, may be relevant if it is something or it reflects something of which the innocent party was, or a reasonable person in his or her position would have been aware, and throws light on the way the alleged repudiatory act would be viewed by such a reasonable person.
4. Although the test for repudiatory breach is simply stated, its application to the facts of a particular case may not always be easy to apply.

References

Case Law

Abram Steamship Co. v Westville Shipping Co. (1923) AC 773
Anns v Merton London Borough Council (1978) AC 728
Arenson v Casson Beckman Rutley & Co (1975) 3 All ER 901
Attwood v Small (1838) 6 Cl & F 232

---

[45] (2010) EWCA Civ 1168. Eminence agreed to sell thirteen flats to Heaney. Thirteen contracts were exchanged in identical terms, each contract providing for the contractual completion date to be fixed by reference to the date on which the relevant flat under construction was ready for occupation. However Eminence's solicitors mistakenly calculated the final date for completion under the notice.

Bell v Lever Brothers Ltd. (1932) AC 161
Bende & Sons, Inc. v Crown Recreation, Inc. 548 F. Supp. 1018, 1022 (E.D.N.Y. 1982)
Bradford Third Equitable Benefit Building Society v Borders (1941) 2 ALL ER 205
British Movietonews Ltd. v London & District Cinemas Ltd. (1952) AC 166
Davis Contractors Ltd. v Fareham Urban District Council (1956) AC 696
Derry v Peek (1889) All ER 1
Dimskal Shipping Co SA v International Transport Workers Federation (The Evia Luck) (1991) 4
     All ER 871
Donoghue v Stevenson (1932) AC 562
Dorset Yacht Co. Ltd. v Home Office (1970) AC 1004
Eminence Property Developments Ltd. v Heaney (2010) EWCA Civ 1168
Esso Petroleum v Mardon (1976) 2 All ER 5
Governors of the Peabody Donation Fund v Sir Lindsay Parkinson & Co Ltd. (1985) AC 210
Great Peace Shipping Ltd. v Tsavliris Salvage (International) Ltd. (The Great Peace) (2002) 4 All
     ER 689
Hadley v Baxendale (1854) 9 Exch 341
Hedley Byrne & Co Ltd. v Heller & Partners Ltd. (1964) AC 465
Horne Coupar v Velletta & Company 2010 BCSC 483
J Lauritzen AS v Wijsmuller BV (The Super Servant Two) (1990) 1 Lloyd's Rep 1
Jacob & Youngs v Kent (1921) 121 NE 889
Keneric Tractor Sales Ltd. v Langille (1987) 2 SCR 440
McAlpine Humberoak v McDermott International (1992) 58 BLR 1
McDonald v Dennys Lascelles Ltd. (1933) 48 CLR 457
McRae v Commonwealth Disposals Commission (1951) 84 CLR 77
Museprime Properties v Adhill Properties (1990) 36 EG 114
North Ocean Shipping Co. Ltd. v Hyundai Construction Co Ltd. (1979) QB 705
Pao On v Lau Yiu (1980) AC 614
Peak Construction (Liverpool) v McKinney Foundations (1970) 1 BLR 114
Photo Production Ltd. v Securicor Transport Ltd. (1980) AC 827
Pioneer Shipping Ltd. v BTP Tioxide Ltd. (The Nema) (1982) AC 724
Radford v De Froberville (1977) 1 WLR 1262
Ranger v Great Western Railway Company (1854) 5 HL Cas 72
Ruxley Electronics v Forsyth (1996) AC 344
Stevenson v Reliance Petroleum Ltd. (1956) S.C.R. 936
Sutcliffe v Thackrah (1974) AC 727
Tamplin v James (1880) 15Ch D 215
Taylor v Caldwell (1863) 3 B & S 826
William Sindall Plc v Cambridgeshire County Council (1994) 3 All ER 932
Woodar Developments v Wimpey (1980) 1 All ER 571

Books

Adams J, Brownsword R (1995) Key issues in contract. Butterworths, London
Allen R, Martin S (2010) Construction law handbook. Aspen Publishers, Maryland
Beatson J, Burrows A, Cartwright A (2010) Anson's law of contract. Oxford University Press,
     Oxford
Beale H (2010) Chitty on contract. Sweet & Maxwell Ltd, UK
Beale H, Bishop W, Furmston M (2008) Contract: cases and materials. Oxford University Press,
     Oxford
Burrows (2005) Remedies for torts and breach of contract. Oxford University Press, Oxford
Furmston M, Cheshire GF, Fifoot CHS (2006) Cheshire, Fifoot and Furmston's law of contract.
     Oxford University Press, Oxford

Harris D, Campbell D, Halson R (2005) Remedies in contract and tort. Cambridge University Press, Cambridge

Hart H, Honore T (1985) Causation in the law. Oxford University Press, Oxford

Jones J (2009) Goff & Jones: the law of restitution. Sweet & Maxwell, UK

Poole J (2010a) Textbook on contract law. Oxford University Press, Oxford

Poole J (2010b) Casebook on contract law. Oxford University Press, Oxford

Ramsey V, Furst S (2008) Keating on building contracts. Sweet & Maxwell, UK

Samuel G (2001) Law of obligations and legal remedies. Cavendish Publications, London

White N (2008) Construction law for managers, architects, and engineers. Delmar Cengage Learning, USA

Chapter 4
Delays and Disruptions Provisions

4.1 Delay and Disruption: An Introduction

Construction and engineering projects are subject to considerable risks and uncertainties. These include weather, soil conditions, availability of labour, materials and plant and sometimes the intervention of certain government bodies and local authorities. Such uncertainties frequently cause delays in project scheduled programmes and, ultimately, the completion of the project. For contractors, and frequently consultants too, these manifest themselves as liquidated or actual damages, labour, material and equipment costs, extended head and site office overheads and loss of productivity costs. For employers, they appear as loss of profit, revenue opportunity costs and consultants' fees. As these costs can be significant, the liability for delays and disruptions is frequently a subject of contention.

Delay and disruption matters that often result in projects finishing late and over budget are often supplemented by enormous claims for compensation or liquidated damages. It is a generally accepted principle of risk management that those who are most able to manage a particular risk should bear that risk, and therefore, the contractor is required to use the tools, available to him, to manage the employer risk and to overcome and avoid unnecessary delay howsoever caused. The contractor will nevertheless need to set out the details of the employer risk events relied upon and the compensation claimed with sufficient particularity so that the employer knows the case that is being made against him. Failing to do so, as is common in many situations, the contractor will produce a global claim with all the causes rolled up in one claim and the losses shown as a lump sum with a gross number showing the difference between the contract actual cost and his bid cost.

Delays and disruptions are two different types of damages; delay damages are valid only if delays to the overall project completion time are involved, while disruption damages can be caused by any change in the planned condition of work that can happen regardless of the change in the project completion time.

A. D. Haidar, *Global Claims in Construction*,
DOI: 10.1007/978-0-85729-730-3_4, © Springer-Verlag London Limited 2011

The expression delay claim is usually used to describe a monetary claim which follows on from a delay to the work as a whole. The expression disruption claim is used to describe a monetary claim in circumstances where part of the works has been disrupted, without affecting the ultimate completion date of the project; this typically equates with delay which is not on the critical path. For the purpose of a delay claim, it is usually taken for granted that the contractor must first establish a right for an extension of time where there is no corresponding presumption in the case of a disruption claim.

Project managers have an interest in risk management, which should extend to knowing the strengths and weaknesses of the project management approaches if a dispute situation arises. Changes to the schedule of works caused by clients and third parties often become contentious issues especially when a change results in delay and disruption. Delay and disruption claims are demonstrated by a logical interpretation of the events. Such claims can be supported by documents, letters, instructions and witness statements, or may extend to the use of computer-aided project management tools. Most project managers, during the course of the contract, are collecting the type of information necessary to enable them to particularise events. While the documentation and recording of such information is primarily for management purposes, it also places them in a good position if they need to pursue a claim. Those failing to collect and document information regarding change will be left with few options. Attempting to collect and assemble data after the event will limit claims to retrospective accounts and analysis.

The tender allowance has limited relevance for the evaluation of delay and disruption caused by breach of contract, or any other cause which requires the evaluation of additional costs. However, the tender allowance may be relevant as a base line for the evaluation of delay and disruption caused by variations.

Time is a complex parameter in the matrix of a construction contract that often leads to delays and disruptions for the concerned project. The doctrines and principles that create causations are:

1. Types of delays, time of essence and the reasonableness test.
2. Extension of time.
3. Completion matters, concurrency, acceleration and *time at large*.
4. Project programming and float. This issue has a complex nature and becoming critical in proving claims and assisting in the calculation of damages.
5. Liquidated damages, the doctrine of *quantum meruit* and their calculations.
6. Mitigation and remedies.

4.2 Types of Delays

A delay is defined as the time during which some part of the construction project is completed beyond the projected completion date or not performed as planned due to an unanticipated circumstance. Delay may be caused not just by the owner or contractor but by any party participating in the project such as the designer, prime contractors, subcontractors, suppliers, labour unions and utility companies. A delay might also result due to *force majeure*,[1] such as a weather, fire or earthquake.

Construction delays can be categorised as two major types namely excusable and non-excusable. An excusable delay is one for which the contractor is excused from meeting a contractual completion date and for which will, therefore, receive a time extension. Excusable delays can be caused by either the owner or a third party not participating directly in the contract. In general, excusable delays include unforeseen design problems, variations and change orders, site restrictions, late payments and Acts of God such as fire, strikes and wars. Even if a delay appears to be excusable, it will be the responsibility of the contractor if it was foreseeable; it could have been prevented but for the acts of the contractor, or it was caused by the negligence of the contractor. This type of delay is also called non-culpable delay.

A non-excusable delay involves lost time caused directly by the contractor actions or inactions. In this case, the contractor is entitled neither to time extension nor to additional compensation from the owner. Moreover, the contractor will be responsible for the possible impact its performance has on other involved parties. If the contract includes a liquidated damage clause, then under this clause the owner could recover delay damages from the contractor. Generally, non-excusable delays include the contractor failure to perform work within the agreed time period, poor work performance and resource availability problems. This type of delay is also called culpable delay.

Excusable delays can be further categorised into two types namely compensable delays and non-compensable delays. A compensable delay allows the contractor both a time extension and additional damage costs. This type of delay is caused by the owners or their representatives such as the consultant, project manager or design team. In this case, the owner should compensate not only for damage costs caused by the compensable delay, but also for the cost of any follow up work necessitated by the delay. The damages also include all costs incurred by the contractor due to the delay such as overhead costs, interest on payments and all related losses such as procurement and design contingencies. In the event of a non-compensable delay, the contractor is not entitled to compensation for additional costs caused by the delay, but may be entitled to a time extension.

Delays can also be classified as critical or non-critical. A critical delay results in the extension of the contract project completion date. Such an event involves the initial delaying of a critical path activity that has zero day of total float, but it will also affect subsequent activities, thereby altering the completion date of the entire project. Conversely, a non-critical delay is either one involving a non-critical path activity that has a float or one that does not extend the contract project completion date. A delay to an area which can be performed at a later stage of the project and

---

[1] Also called Act of God.

has a positive float will normally not affect the project completion date and, therefore, is noncritical in relation to the overall completion date. Commonly, a non-critical delay is a delay for which the contractor is not entitled to a time extension, but may actually recover some damages as it can cause disruption to his overall performance.

4.3 Disruption Implications

Disruption is often treated by the construction industry as if it were the same thing as delay. They are, however, two separate matters. Delay is lateness, whereas disruption is loss of productivity, disturbance, hindrance or interruption to the contractor normal working methods, resulting in lower efficiency. Disrupted work is work carried out less efficiently than it would have not been for the cause of the disruption. Disruption compensation is only recoverable to the extent that the employer caused the disruption. Most standard form of contracts do not deal expressly with disruption, but disruption may be claimed as a breach of the term generally implied into construction contracts that the employer will not prevent or hinder the contractor in the execution of its work.

Disruption costs may be distinguished from delay costs by virtue of the fact that the latter are a function of time and the former are essentially productivity related. In a disruption claim, contractors claim that they could not achieve their planned output, because of the employer actions or other causes not their responsibility, and hence that the damages or extra costs are payable.

Causes of disruption can be broken down into either external or internal causes. External causes of disruption are generally not related to the project itself and will often fall into the *force majeure* category. They will also include government acts such as the passing of new regulations, changes to taxes and new laws. Such events will generally involve certain rights of compensation to be passed on to the contractor. External causes can also result from Acts of God, such as earthquakes or drastic weather conditions. Such risks can generally be insured against, and therefore the uncertainty that they would introduce into a project can be transformed into a certain insurance cost.

The internal causes of disruption can be causally attributed to the project itself, its planning and design and the manner in which the works are performed. Internal causes of disruption can be further broken down into: firstly technical causes including changes in design, design errors and construction errors; and secondly economic causes including difficulties in accessing requisite materials, labour or skills and financial causes such as lack of client funds, material or labour cost increases and interest rate rises.

It will usually be accepted on a complex project that a certain amount of uncertainty and rework will be expected at various stages of the construction. Even when the project is going well, normal disruptions, made by both the contractor and the client, will involve a certain amount of rescheduling and planning and additional costs to rectify. Despite the fact that these costs are built into the initial tender and will be absorbed without affecting the time frame or budget, it is possible to drastically underestimate the costs of such factors.

There are other types of disruptions that can be significant in their impact and are rarely thought about during the original estimating. When these types of disruption do occur, their consequences can be underestimated as they are often seen by the contractor as damages incurred and should be compensated by the client. One of the most common causes of these types of disruption is a variation or change order, coming from the client, and amending what the contractor is required to do, or what the project is required to deliver. This can occur even after the work has commenced.

To justify a disruption claim, a contractor must establish that actual progress of the work has been interrupted and the cause of the disruption was either a breach of the contract by the employer or an action for which the contract provides for the reimbursement of extra cost. Even if the client does cause disruption, this may not result in an entitlement to additional payment. It may be that the contractor failed to comply with certain contractual requirements and is therefore not entitled to reimbursement of the disruption costs. A standard normal requirement for the validity of a disruption claim is that the contractor must give notice of information required at the time of knowing of the disruption. The contract usually will provide the mechanism for the notification as the type of information required and to whom the notice should be addressed.

The practices, which are determinant to success and failure of entitlement to contractor compensation due to disruption, are summarised:

1. The work that has been affected must be clearly identified, and the work activities that were affected by the disruption must be specified. The extra expense incurred must be explained.
2. The contractor must show that the event leading to the disruption and financial loss was either a breach of contract, or an event provided for in the contract for which the employer is to be made financially liable to the contractor.
3. It must be shown that actual work progress has been negatively impacted. It is not sufficient to show that planned future work has been impacted as such uncertainties may never materialise.
4. The contractor must quantify the disruption costs using a selected method of quantification. The principle that applies is that the extra costs incurred, compared to the costs that would have occurred had the disruption not occurred, are recoverable by the contractor.
5. The contractor sets out what the actual costs would have been had the disruption not occurred. This provides the base line to be used in the calculation.
6. The contractor must show that he has taken all reasonable steps to mitigate his loss, such as returning leased equipment, working on other parts of the project that were not affected by the disruption and redeploying expensive resources so that they are not unnecessarily sitting idle.

4.4 Time of Essence

Where the contract has no express provisions as to time for completion of the works, the contractor is obliged to complete the works within a reasonable time. What is reasonable depends upon both what was anticipated at the outset, including the anticipated level of resources, and matters which occur during the project over which the contractor had no control.

Generally, the parties would have agreed a date for completion which will be binding. The date for completion ceases to apply if there is a delay caused by a breach of contract or an act of prevention by the employer and there is no applicable provision in the contract for an extension of time.

In such circumstances, when there is no relevant stated date for completion, the obligation to complete reverts to being one to complete within a reasonable time. This may not, however, mean the same as a reasonable time referred to in above, but it may be more a matter of adding to the agreed period additional time to reflect the impact of the breaches of contract or acts of prevention. A second consequence of there being no applicable fixed date for completion is that any liquidated damage provision becomes ineffective with the result that the employer entitlement to damages, for delay beyond the reasonable period, is compensation for losses which he can prove.

There is no general concept of time being of the essence of a contract as a whole. Instead, the question is whether time is of essence of an individual term. It is due to the nature of construction work, where the works become fixed to the employer land and cannot readily be removed, time is not being considered to be of the essence unless the parties expressly stipulated that conditions as to time must be strictly complied with and the nature of the subject matter of the contract or the surrounding circumstances show that time should be considered to be of the essence.

Time is of the essence clause generally carries far less weight in construction contracts than other commercial contracts. The reasons include the fact that construction contracts typically contain internal remedies addressing delay and the disproportionate effect of the remedy as compared to the breach, particularly when a building contract is partially preformed.

Construction contracts involve various stages of development, numerous parties and countless variables. Additionally, delay is often consequential, expected and outside the control of either party. Moreover, most construction contracts incorporate a variety of terms compelling the contractor to perform its duties in a timely fashion such as liquidated damages and express termination provisions specifically addressing delays in performance. These specific clauses may well override a generic clause declaring time to be of the essence as they raise the question as to whether the parties intended the clause to operate in a field occupied by an express provision. Indeed, there is a good argument that where a party stipulates for liquidated damages; it has declared an intention that damages are an adequate remedy, meaning the time obligation is not a condition that would entitle that party to terminate the contract.

Another difficulty, in giving effect to time is of the essence clauses in construction contracts, is the sheer number of time reference in construction contracts for various duties, obligations and notices. A missed time deadline in a construction contract may well arise after substantial performance leading to concerns of unjust enrichment. Accordingly, there may be reluctance in arriving at an interpretation that permits termination.

This general approach of the notion as of time is of essence in construction contracts and the reasoning behind that approach is reflected in *Hudson's Building and Engineering Contracts*:

> However, in examining a contractor's obligation to complete his work to time, construction contracts differ very markedly from nearly all others in that the contractor can be expected to have expended very heavily in performing the contract prior, for example, to a relatively trivial delay after completion, and also that upon fixing of the work to the soil the property in it will have passed to the owner irrespective of the degree of payment, thus conferring a major and irretrievable benefit on the owner as against a possibly only minor or nominal loss suffered by him. No doubt for these reasons the courts have shown an exceptional assiduidity in avoiding a time of the essence interpretation of the contractor's completion obligation in construction contracts, it would seem even in cases where express language has been used in the contract.

Analysis of construction contracts in common law jurisdictions is rooted in the fundamental principles of contract law. At common law, a contract which specifies the time of performance is normally regarded as of essence. Failure to perform a stipulation as to time does not differ intrinsically from any other failure to perform. The attitude of the courts, to the principles of time of essence in construction contracts, is to identify the machinery type provisions rather than strict, essential time conditions.

In *Mount Charlotte Investments Ltd v Westbourne Building Society*,[2] J. Templeman sets, *obiter dicta*, the three conditions that make time of essence as follows:

- The contract expressly stating that this is so.
- Implication because of the special matter (for example, the completion of the project will allow the kick start of another).
- Notice from the innocent party making time of the essence after the other party has defaulted under the clause.

In *Raineri v Miles*[3] the House of Lords confirmed the same as above to the matter of time of essence and that "failure to adhere to the timetable was not a breach of the contract." According to Lord Fraser of Tullybelton:

> The principle which in my opinion emerges from the authorities to which I have referred is that breach of a contractual stipulation as to time which is not of the essence of a contract will not be treated as breach of a condition precedent to the contract, that is as a breach which would entitle the innocent party to treat the contract as terminated or which would

---

[2] (1976) 1 All ER 890.

[3] (1980) 2 All ER 145.

prevent the defaulting party from suing for specific performance. Nevertheless it is a breach of the contract and entitles the injured party to damages if he has suffered damage.

In view of the above said, the insertion of a clause declaring time to be of the essence in a construction contract, unlike its insertion in other contract forms, will not normally, in and of itself, allow the innocent party to rescind or terminate the contract for any breach of a time condition. In determining the party intentions, the court will look to all the particular terms and circumstances and may well import little meaning to the time is of the essence clause and therefore construction contract drafters should be aware of the potential limitations of a time is of the essence clause and give some thought to how they use such clauses.

4.5 Reasonable Time

An obligation to complete within a reasonable time arises either because the contract is silent as to time, or because the specified time has ceased to be applicable by reason of some matter for which the employer is responsible. What is a reasonable time may not depend solely upon the convenience and financial interests of the contractors. Where the time for completion is not given in the contract documents, the law provides for a term to be implied that the work will be carried out within a reasonable time.

Reasonable time is primarily a question of fact and must depend on all the circumstances which might be expected to affect the progress of the works. In calculation of a reasonable time, all the circumstances of the case should be taken into consideration, such as the nature of the works to be done, the time necessary to do the work, the ability of the contractor to perform and the time which a reasonably diligent contractor would take to perform a similar task with similar constraints.

At the turn of the century, the general rule of law was that any act necessary to be done by either party in order to carry out a contract must be done within a reasonable time. The principle was interpreted that, except where it is stipulated that time is of the essence, a breach of contract was only committed in the case of unreasonable delay in the performance of any act agreed to be done. However, by the middle of the century, it was found that a breach of a contractual stipulation as to time, which is not of the essence of a contract, would not be treated as a breach of a condition precedent to the contract and therefore would entitle the innocent party to treat the contract as terminated or would prevent the defaulting party from suing for specific performance.

What is a reasonable time was considered by J. Diplock in *Neodox Limited v The Borough of Swinton and Pendlebury*.[4] This case involved the question of whether the engineer acting for the defendant corporation had failed to issue

---

[4] (1958) 5 BLR 38.

instructions to the contractors within a reasonable time. J. Diplock explained what was meant by that expression in this way:

> In determining what is a reasonable time as respects any particular details and instructions, factors which must obviously be borne in mind are such matters as the order in which the engineer has determined the works shall be carried out, whether requests for particular details or instructions have been made by the contractors, whether the instructions relate to a variation of the contract which the engineer is entitled to make from time to time during the execution of the contract, or whether they relate to part of the original works, and also the time, including any extension of time, within which the contractors are contractually bound to complete the works.

J. Diplock ruled that what was a reasonable time did not depend solely on the convenience and financial interests of the contractor. He observed that while it may appear to the contractor that it is in his interest "to have every detail cut and dried on the day the contract is signed", such a state of affairs could not have been contemplated at the time of the contract. He then proceeded to hold that what was a reasonable time was a question of fact to be determined with reference to all the circumstances of the case. These include:

1. Considerations of the employer's engineer and his staff.
2. The order by which the works were to be carried out and approved by the engineer.
3. The contractor's requests for particular details.
4. Whether the details requested relate to variations.
5. The length of the contract period.

Admittedly the list of factors, tendered by the *Neodox* judgment, cannot be considered to be exhaustive. However, it does serve to indicate the wide range of factors that has to be considered when determining the question of reasonableness with respect to the timing of instructions, additional drawings and information.

In *British Steel Corporation v Cleveland Bridge & Eng Co.*,[5] it was decided that to complete within a reasonable time is an implied term. What constitutes a reasonable time has to be considered in relation to circumstances which existed at the time when the contract obligations are performed, but excluding circumstances which were under the control of the contractor. Lord Goff applied these principles by first considering what in ordinary circumstances a reasonable time for performance is and then considering to what extent the time for performance of the contractor is in fact extended by extraordinary circumstances outside his control. Whether a reasonable time has been taken to do the works cannot be decided in advance, but only after the work has been done.

---

[5] (1981) 24BLR100.

4.6 Extension of Time

The benefit to the contractor for an extension of time is only to relieve the contractor of liability for damages for delay, usually liquidated damages, for any period prior to the extended contract completion date. The benefit of an extension of time for the employer is that it establishes a new contract completion date and prevents time for completion of the works becoming at large. Extension of time notices should be made and dealt with as close in time as possible to the delay event that gives rise to the application. Some delays are related to time extension, but some are not. In a broad view, much depends upon the party responsible for the delays and when the delays occur.

Standard forms of contract often provide that some kinds of delay events, adverse weather conditions and strikes being common examples, which are at the risk of the employer so far as time for completion is concerned carry no entitlement to compensation for delay. They are sometimes misleadingly called neutral events as they are only neutral in the sense that one party bears the time risk and the other party bears the cost risk.

Beyond the matter of a breach or an act of prevention, the general law gives no protection to a contractor who is delayed in carrying out the works except in the extreme cases where the requirement of the doctrines of frustration, misrepresentation or duress is satisfied. However, to achieve a more balanced allocation of the risk of delaying events, it is common practice for construction contracts to include express provisions for the granting of extensions of time for completion of the works.

The inclusion of extension of time clauses within any construction contract is thus advantageous to the contractor and the employer alike as extension of times enable the completion date to be adjusted in the event of certain causes of delay not caused by the employer. To this extent, the extension of time provision protects the contractor against liability for delay. The extension of time also operates to the employer benefit in certain circumstances. For example, where the contractor is delayed due to the employer default and there is no express provision for an extension of time within the contract, the original completion date simply falls away and the principle of time at large applies.

It is not generally the contractor duty to ascertain the reason for an extension of time, rather, the ground to an extension of time should be granted is a matter of opinion that the contract manager bases on his own decision. In some contracts a period for responding to an application for an extension of time is stated and in others the time is left open and a requirement for a decision within a reasonable time is implied. The principle steps leading to an extension of time are summarised as follows:

- When it becomes apparent that the progress of the works is being or is likely to be delayed then the contractor shall notify the architect or the project manager of the cause of delay and identify whether in his opinion it is a relevant event.

- The contractor is required to provide with the notice, or as soon as possible after the notice, particulars of the expected affects of the event and an estimate if any of the expected delay to the completion of the works beyond the completion date.
- Upon the receipt of the notice and any further particulars the architect is required to decide whether in his opinion any of the events notified are relevant events and whether as a result of such events the works are likely to be delayed beyond the completion date. If he so decides, then he is then required to give an extension of time to the contractor in writing as he then estimates to be fair and reasonable.

There is a great deal of case law about the meaning of a particular ground for an extension of time under specific contract forms. There are also some points of general application to be noted in relation to what constitutes a sufficient extension of time provision for the employer default. In *Tan Ah Kian v Haji Hasnan*,[6] J. Gill identifies, *obiter dicta*, the following situations when an extension of time can be granted:

- The wording must be reasonably specific in showing that it is intended to cover the employer defaults.
- There is authority suggesting that an extension of time for a breach or an act of prevention may only be granted prior to the completion date unless there are clear words allowing it to be granted afterwards.
- There is debate over whether the existence of a procedural requirement which takes the form of a condition precedent to an entitlement to an extension of time may have the effect that, if not satisfied, there is said to be no effective extension of time provision so that time is set at large.

In *Peak v McKinney*,[7] it was held that, to entitle the employer to grant extra time for his own defaults, the extension of time clause must provide, expressly or by necessary inference, for an extension on account such a fault of breach is on the part of the employer. According to LJ Salmon:

> No doubt if the extension of time clause provided for a postponement of the completion date on account of delay caused by some breach or fault on the part of the employer the position would be different. In such a case the architect would extend the date for completion, and the contractor would then be liable to pay liquidated damages for delay as from the extended completion date.

---

[6] (1962) MLJ 400.

[7] Peak Construction (Liverpool) Ltd v McKinney Foundations Ltd (1970) 1 BLR 111. This is a case where the works were suspended due to defective piles for which the contractor was responsible, but the employer caused further delay for which there was no mechanism in the contract for extending time. The main contract allowed the architect to certify extensions of time for additions to the works, strikes, *force majeure* or any other unavoidable circumstances. These provisions did not permit an xtension of time to be granted for the employer failure to promptly authorise and instruct the investigations and remedial works.

Often, the extension of time clause stipulates that the extension of time should be fair and apply the rules of the contract. What constitutes the basis of a fair and reasonable extension of time was considered in *John Barker Construction Limited v London Portman Hotel Limited*, [8] where HHJ Toulson QC set out the criteria that must apply in order to calculate a fair and reasonable extension of time which comprises the fact that the contract manager recognises the effects of constructive change, makes a logical analysis, in a methodical way, of the effect of relevant events on the contractor programme and to calculate, rather than to make an impressionistic assessment of, the time taken up by relevant events. These criteria should be regarded as a form of guidance as to one approach, but not universally binding, because the wording of the contract is what matters most.

4.7 Gross and Net Methods

There are two schools of thought or methods on how the extension of time should be calculated where an extension of time is granted during a period of delay. The first method, described as the gross method, has been preferred by many academics and some commentators and propounds that if an extension of time is granted because of an event arising during a period of culpable delay, which is a delay wholly the responsibility of the contractor, then the extension of time must begin to run from the date the event occurred was given. This means that the architect must establish a new completion date for the contract which adds the extension of time from the date of the instruction, thus denying the employer liquidated damages up to the new completion date.

This has traditionally been a contractor led argument. It arises out of the case of *Wells v Army & Navy Cooperative Society*.[9] In that case, under a building contract, in the execution of which there had been substantial delay, and which involved a provision for liquidated damages, certain matters causing delay and other causes beyond the contractor's control were to be submitted to the board of directors of the owners of the building who were to 'adjudicate thereon and make due allowance therefore if necessary, and their decision shall be final'. The drift of the approach appears from the words of J. Wright:

> Some of the details were not even supplied until after the expiration of the time for completion. The only answer given by the architect is that in his view the plaintiffs were not ready to go on with the work for which the details were asked. I think that this, even if proved, is not a sufficient answer. The plaintiffs must within reasonable limits be allowed to decide for themselves at what time they are to be supplied with details. It is very difficult to determine how far any particular defaults of this kind on the part of the

---

[8] (1996) 83 BLR 31.

[9] (1902) 86 LT 764. The Wells case is curiously relevant in another context, namely that it is an early forerunner of the prevention arguments which appeared much later and with significantly more force in cases like *Peak v McKinney* (1970) (*supra*).

defendants would entitle the plaintiffs to relief from penalties, especially when, as in this case, there were other and more important causes of delay which would not be grounds of relief in this action; but on the whole I think that the delays in giving details not merely contributed to the delay of completion, but were such as even in the absence of the other causes of delay would have prevented completion in due time, and in my view to a great extent increased the delay of completion.

The second method, known as the net method of calculation, has been more favourable with the courts. The net method is where the contractor is only entitled to an extension of time equal to the time required to carry out the additional work. Effectively, if the contractor is six months in delay and is delayed by one further month due to a relevant event, then the completion date would be extended from the original completion date to a month later, still leaving the contractor with five months of culpable delay and the threat of liquidated damages.

Lord Denning explained, in the case of *Amalgamated Building Contractors Ltd v Waltham Holy Cross UDC*,[10] that the power of extending contractual completion dates should apply retrospectively and that common sense requires that the method of assessment of such an extension would be what he termed the net method. Lord Denning stated:

> Take a simple case where the contractors, near the end of the work, have overrun the contract time for six months without legitimate excuse. They cannot get an extension for the period. Now suppose that the works are still uncompleted and a strike occurs and lasts a month. The contractors can get an extension of time for that month. The architect can clearly issue a certificate which will operate retrospectively. He extends the time by one month from the original completion date, and the extended time will obviously be a date which is already past.

In *Balfour Building v Chestermount Properties*,[11] the court confirmed that the purpose of the power to grant an extension of time was to fix the period of time by which the period available for completion ought to be extended having regard to the incidence of relevant events. The completion date, as adjusted, was not the date by which the contractor ought to have achieved practical completion, but the end of the total number of working days starting from the date of possession, within which the contractor ought fairly and reasonably to have completed the works. On this footing, J. Coleman, *obiter dicta*, clarified the issue in regards to where a relevant event arose after the date for completion and during a period in which the contractor was in culpable delay, the contractor would only become entitled to a net extension of time corresponding to the specific number of days of delay occasioned by the relevant event. In other words, the occurrence of the new delaying event would let the contractor liable for its own culpable delays. He concluded:

> Accordingly, I conclude on the second question that it would be wrong in principle to apply the 'gross' method, and that the 'net' method represents the correct approach.

---

[10] (1952) 2 All ER 452 CA.

[11] Balfour Building Ltd v Chestermount Properties Ltd (1993) 62 BLR 1 QBD.

4.8 Completion of the Works

Commencing, proceeding and completing the works are three major components in any construction project where time factors are essential in interconnecting the different variables affecting the process of executing construction in a timely manner. The relationship between these factors is complex and creates a web of interactive activities which should be identified before the start of the works and updated constantly during the project.

In the absence of an express term there may be an implied term in every construction contract that the contractor shall commence the works within a reasonable time of making the contract and proceed with the works at a reasonable rate. The completion of the works comprised within a contract is critical as not only it has a direct bearing on the question of whether the employer can levy liquidated damages on the contractor, but it also usually marks the transfer of certain risks or the crystallisation of certain rights between the contractor and the employer *inter se*.

A reasonable time for completion will usually be implied when the contract fails to specify the time for completion of the works. In the absence of an express provision as to progress, business efficacy may also require the implication of a term that the contractor will proceed with reasonable diligence and maintain reasonable progress during the contract period.

However, if the contract provides for completion of parts of the works by key dates and for damages and forfeiture for failure to execute the works with due diligence and expedition, there is no implied term as to the implication of a requirement to execute the works with due diligence and expedition but only a duty to proceed with such diligence and expedition as were reasonably required in order to meet the key dates and the completion date in the contract. Accordingly, the contractor has the freedom to plan his work as he sees fit within the specified time constrains and is not obliged to proceed at a particular work rate. He must ensure, however, that progress is not so slow that he can be said to be deliberately putting himself in a position where he cannot complete on time, which could amount to an anticipatory repudiatory breach of contract.

In *Wells v Army and Navy Co-operative Society* (1902) (*supra*), LJ Williams defines completion by stating:

> In the contract one find the time limited within which the build to do this work. That means not only that he is to do it within that time, but it means that he is to have that time within which to do it. To my mind that limitation of time is clearly intended, not only as an obligation, but as a benefit to the builder. In my judgment, where you have a time clause and a penalty clause, it is always implied in such clause that the penalties are only to apply if the builder has, as far as the building owner is concerned and his conduct is concerned, that time accorded to him for the execution of the works which the contract contemplates he should have.

There have been many shades of opinion as to what constitute completion and the use of some phrases in certain standard forms of building and construction contracts such as practical completion and substantial completion. In *J. Jarvis and Sons v Westminster Corporation*,[12] LJ Salmon defined practical completion as completion for the purpose of allowing the employer to take possession of the works and use them as intended. He, therefore, held that practical completion did not mean completion down to the last minute details. He stated the following:

> We would normally say that a task was practically completed when it was almost but not entirely finished; but practical completion suggests that is not the intended meaning and what is meant is the completion of all the construction work that has to be done.

In *H W Neville v William Press & Sons*,[13] it was held that practical completion was achieved when the works comprised in the contract had been carried out, save for minor works, and if there were any patent defects, the architect should not certify practical completion. The court held that the word practical completion gave the architect a discretion to certify that the contractor had fulfilled its obligation, even though very minor, *de-minimis*, work had not been carried out, but that if there were any patent defects in what the contractor had done the architect could not have given a certificate of practical completion.

In *Emson Eastern Ltd v E.M.E. Development Ltd*,[14] J. Newey, in arriving at a decision, took account of what happens on building sites. He concluded that there was no difference in the meaning between completion and practical completion. Completion, he considered, was like practical completion, something which occurs before defects and other faults have to be remedied. He stated:

> Because a building can seldom if ever be built precisely as required by drawings and specification, the contract realistically refers to practical completion and not completion, but they mean the same. If, contrary to my view, completion is something which occurs only after all defects, shrinkages and other faults have been remedied and a certificate to that effect has been given, it would make the liquidated damages provision unworkable.

Notwithstanding the definitions referred to above by the different authorities, it is often the case that individual circumstances can and do affect the issue of a certificate of practical completion, irrespective of the condition of the works on site. Such circumstances may include the client willingness to accept the works which may be dependent upon whether he has issued a lease to occupy them by a certain date and therefore accept the cost of maintaining, securing and insuring the works when not completely finished and free of defects.

---

[12] (1969) 1 WLR 1448 CA.

[13] H W Neville (Sunblest Ltd) v William Press and Son Ltd. (1981) 20 BLR 78.

[14] (1991) 55 BLR114. Practical completion was certified but some time after Emson went into administrative receivership and his employment was automatically determined. The issue was whether Emson were entitled to further payment. The matter turned on whether completion meant the same as practical completion, or whether it meant that all snagging and remedial works has to be made good at the end of the defects period before the works could be said to be complete.

Given the ambiguity in current case law and the currency at which this matter arises, it is surprising that the more frequently used standard forms of contract contain no clearer definition for practical completion. In the absence of such clear definition or persuasive case law either way, it is at the mercy of the client and the architect and their interpretation of practical completion that the contractor has to argue his position accordingly.

4.9 Programming of the Works

Delays have been found to be the most cited source of disputes and the most costly cause of problems on construction projects in many contractual regimes. Given this state of affair, it is also noticeable that cases that have come before the court, where time disputes involving delay and causation issues are central to the proceedings, rarely involve the use of programming techniques as a method of reliable analysis. Therefore, the courts are increasingly demanding clearer explanations of cause and effect and in complex construction projects detailed time impact analysis.

Project programming consists of updating current and target schedules for existing projects and developing breakdown structures, milestones, target schedules and cost-loaded schedules for new projects. The resolution of disputes on large construction and engineering contracts increasingly involves the use of delay analysis techniques to assist in the identification of the cause of critical delay to a project and to assist in the computation of claims for lost productivity. While the industry is becoming more and more familiar with the use of the tools and techniques employed in the process of delay analysis, unfortunately at present, there is very little common agreement upon their correct application and understanding.

As the complexity of projects and the requirements for scheduling have increased, the opportunities for delay to the various activities which have been scheduled and are necessary for the completion of the project. In fact, even determining whether completion of the total project or a phase of the project has been delayed can be a difficult analytical task.

Since delay usually leads to cost increases, there is a need to correctly determine the allocation of the delays, the responsibility of the parties and causation. With this allocation, there can be a technically sound foundation for an acceptable resolution of delays cost attribution. With the increasing use of the programming methods the process of analysing and identifying the varying situations is facilitated.

Most construction contracts contain provisions for the contractor to submit to the employer a programme showing the manner in which the contractor intends to carry out and complete the works. This requirement may range from a simple request for the submission of a master programme without prescription as to its form or content, or in some cases, a very detailed requisition as to the format, content and operation of the programme to be submitted by the contractor.

In whatever form it is submitted, the programme is a crucial document for the effective management of most construction projects as it provides a tool by which actual job progress against a plan is monitored, thus, enabling an early alert of actual and potential delays which could adversely affect the project completion date.

In terms of the programme use during the contract and post-contract period, it will not necessarily establish the impact and extent of delay unless more sophisticated forms of analysis are adopted and are used on an ongoing basis throughout the project so that the programme becomes literally a working and living programme. This is because if delay allegations are to be shown effectively by the contractor and considered properly by the architect or the engineer, it will be found that in most situations that a simple bar chart will not suffice and some better means of indicating quantity output or physical progress, as well as the passage of time is essential.

In *John Barker Construction Ltd v London Portman Hotel Ltd* (1996) (*supra*), the way and manner contractors delay and disruption claims assessment should be carried out was a fundamental issue. In this case, it has been established that in exercising his duty, the architect or contract administrator must undertake a logical analysis in a methodological way of the impact of the relevant events on the contractor programme and that the application of an impressionistic rather than a calculated and rational assessment is not sufficient. J. Toulson QC stated that:

> I accept that Mr. Miller believed, and believes, that he made a fair assessment of the extension of time due to the plaintiffs. It is fairly apparent that the defendants were concerned by the overrun of the contract in time and costs, and I have no doubt that Mr. Miller was conscious of this, but I believe also that he endeavoured to exercise his judgement independently. However, in my judgment his assessment of the extension of time due to the plaintiffs was fundamentally flawed in a number of respects, namely: (1) Mr. Miller did not carry out a logical analysis in a methodical way of the impact which the relevant matters had or were likely to have on the plaintiffs' planned programme. (2) He made an impressionistic, rather than a calculated, assessment of the time which he thought was reasonable for the various items individually and overall.

The price of failing to establish a proper programme is well illustrated by the case of *Balfour Beatty Construction Limited v The Mayor and Burgess of the London Borough of Lambeth*.[15] The case concerned a challenge to the enforcement of an adjudicator decision on the basis that the adjudicator prepared his own collapsed as-built analysis in the absence of a delay analysis from the referring party, and reached his decision without giving the responding party an opportunity to comment on his methodology. In reaching his decision, J. Lloyd observed that:

> Despite the fact that the dispute concerned a multi-million pound refurbishment contract, no attempt was made to provide any critical path. It seems that Balfour Beatty had not prepared or maintained a proper programme during the execution of the works. By now, one would have thought that it was well understood that, on a contract of this kind, in order to attack, on the facts, a clause 24 certificate for non-completion (or an extension of time determined under clause 25), the foundation must be the original programme (if capable of

---

[15] (2002) 1 BLR 288.

justification and substantiation to show its validity and reliability as a contractual starting point) and its success will similarly depend on the soundness of its revisions on the occurrence of every event.

Proper planning can save time, and agreement on programmes between client and contractor, at an early stage, can make dispute resolution much less contentious. As the courts become more familiar with delay analysis techniques, it is likely that there will be an increasing number of reported cases addressing these issues giving guidance to delay analysts as to the preferred approaches to take and censuring experts who fail to present cogent and balanced evidence that assists the court in their decision-making.

4.10 Critical Path Method Applications

One of the primary, if not the single most important, methods for project programming involve the critical path method scheduling techniques, which have been the most accepted in the industry because they serve not only as a tool for planning projects, but also as a means to minimise time-related claims, to justify actual claims and to assist in negotiating timely solutions of both in and out of court settlements.

The critical path method is a tool that demonstrates the shortest possible path to completion at any stage by breaking down the interrelationship of the discrete elements that comprise the activities to be undertaken. It is a mathematical and logical tool that can be used to predict the duration it will take to complete a series of activities. A well constructed programme using the critical path method allows the parties to identify which parts of the projects, called activities, are critical.

The longest path of the resulting schedule is called the critical path. It consists of activities that, if delayed, will extend the project beyond its predetermined completion date. In addition to the critical path, there are other various side paths called non-critical paths. If affected by improper scheduling or performance delays, these paths could become critical and thus alter the original critical path. A critical path method is designed to advise involved parties about the relative importance of performing certain activities within the project completion parameters. It indicates to participants whether their work is critical, non-critical, or has any float[16] associated with its performance.

The advantages of using a critical path method can be summarised as follows:

---

[16] Float is the amount of time by which an activity can be delayed without delaying the completion date of the entire project.

- They require detailed analysis of the project, and therefore the scheduler or project manager in charge of the programming would have a better understanding of the project. This requirement minimises the possibility of erroneous or misleading schedules.
- They are well established and easy to understand, with techniques for drawing and calculating project durations developed from advanced high technologies. The critical path method, like any other standard quantitative tool, is widely used and accepted by the construction industry.
- The critical path method can be used to determine the length of delays or additional times needed as a result of unexpected events occurring or changes demanded during the construction process.

However, when used in delay analysis, the critical path method schedules should be realistic and reasonable. The accuracy of duration estimates in the computation procedure for a typical critical path method depends on many factors including:

1. Methods of construction.
2. Resource availability.
3. Work quantity.
4. Nature or complexity of work.
5. Labour and equipment productivity.
6. Quality of filed management.
7. Weather and site conditions.
8. Concurrent activities.

Review boards and courts have accepted the use of the critical path method to prove delay, identify the causes of such delays and inefficiencies and assign responsibility for them. J. Dyson in *Henry Boot v Malmaison*[17] confirmed the importance of the critical path method in establishing causation and that a delay is relevant when it falls on the critical path. He stated the following: "The respondent was entitled to respond to the claim both by arguing that the variations, late information and so on relied on by the claimant did not cause any delay because they were not on the critical path and positively by arguing that the true cause of delay was other matters."

This principle also received support in *Brompton v Hammond*,[18] where J. Seymour confirmed, *obiter dicta*, that in determining a fair and reasonable extension of time as a consequence of a delay event, an examination of the actual critical path of the contractor works should be carried out to establish that the delay event affected, or was likely to affect, the completion of the works. Furthermore, he emphasised that the work activities that were critical to the progress

---

[17] Henry Boot Construction (UK) Ltd v Malmaison Hotel (Manchester) Ltd (2000) CILL 1572 TCC.

[18] The Royal Brompton Hospital v Hammond & Others (No. 7) (2001) 76 Con L.R. 148.

of the works at the time the delay event occurred should be taken into account. J. Seymour made the following comment:

> In order to make an assessment of whether a particular occurrence has affected the ulti-mate completion of the work, rather than just a particular operation it is desirable to consider what operations, at the time of event with one is concerned happens are critical to the forward progress of the work as a whole.

These two cases, *Henry Boot v Malmaison* (2000) (*supra*) and *Brompton v Hammond* (2001) (*supra*), were reconsidered in the case of *Motherwell Bridge Construction v Micafil*,[19] where J. Toulmin commented the following:

> Crucial questions are (a) is the delay in the critical path? and if so, (b) is it caused by Motherwell Bridge? If the answer to the first question is yes and the second question is no, then I must assess how many additional working days should be included. Other delays caused by Motherwell Bridge (if proved) are not relevant, since the overall time allowed for under the contract may well include the need to carry out remedial works or other contingencies.

J. Toulmin went on to add that the approach must always be tested against an overall requirement that the result accords with common sense and fairness. In order to show that an event was not on the critical path, the defendant has to argue that the claimant version of the critical path is incorrect and must prove on the balance of probabilities that the critical path went elsewhere.

US construction case law dominates the references to the critical path method citations as judges showed a willingness to not only understanding the mechanism of a critical path method but to other programming issues such the processes commonly used to prove cause and effect in delay-related disputes. In *Natkin & Co v George A. Fuller Co*,[20] one of the findings adopted by the judge in order to reach his decision was that "the critical path plan may become obsolete unless it is kept current." The judge emphasised that a critical path method schedule usefulness as a barometer for measuring time extensions and delay damages is necessarily cir-cumscribed by the extent to which it is employed in an accurate and consistent manner to comport with the events actually occurring on the job. Furthermore, updating the critical path method during the life of the project is incremental and without doing so can make the schedule redundant.

---

[19] Motherwell Bridge Construction Ltd (T/A Motherwell Storage Tanks) v (1) Micafil Vakkuumtechnik Ag (2) Micafil Ag (2002) 8 Con LR 44. This was a long judgement dealing with many issues, but one of the most significant aspects of this case is that it supports the view that when a contractor has to do additional work, but it nevertheless required to complete by a date earlier than the delay arising from the additional work, then the contractor should be entitled to recover his additional costs in achieving the earlier date, which in this case were the cost of working night shifts.

[20] 347 F. Supp. 17 (W.D. Mo. 1972).

In *Fortec Constructors v United States*,[21] the contractor sought time extensions, extra costs and liquidated damages for a government project to build fuel maintenance facility on an air force base. The court rejected a critical path method prepared by the US government to show that the extra work was not on the critical path and, therefore, the contractor was not entitled for any time extensions. The reason given was that "if the critical path method is to be used to evaluate delay on the project, it must be kept current and must reflect delays as they occur."

In *John Driggs Company, Inc.*,[22] the contractor sought additional compensation and extensions in contract time for multiple events that occurred early in the contract. The board commented that a common thread running through all of these alleged delays is that Driggs did not complete these particular tasks on the originally planned and scheduled date and, therefore, when a significant owner caused construction delay occurs, the contractor is not necessarily required to conduct all of his other construction activities exactly according to his original schedule and without regard to the changed circumstances resulting from the delay.

4.11   Critical Path Method Weaknesses

It is important to recognise that it is easy to manipulate a critical path method in order to derive the required end result. Planners and programmers tend, in certain situations, to alter the sequence and the scheduling mechanism to create preferred results for the party construing it. For example, if a programmer wishes to make a certain section of the work critical, he achieves this by fixing durations of activities or logic links between activities. Equally, if there had been variations issued in part of the works, it is possible to make this element of the programme critical. There is also a tendency for those involved in preparing the critical path method to get lost in the analysis rather than focusing on the important task of establishing the entitlement arising from the events to be modelled.

Courts have occasionally had difficulty in understanding and accepting the logic behind a critical path method and other programming issues such as the processes commonly used to prove cause and effect in delay-related disputes. This may be due to a distrust or lack of understanding of the techniques as there are clearly a number of important questions over their probative value since they require often an extensive matrix of facts in the form of planned and as-built dates to be established and the basis of these facts must be understood given that these dates will often be an important factor in determining whether an event or activity is on or off a critical path. In other words, there is commonly a degree of interpretation as to what defines completion and when an activity actually started or finished.

---

[21]   8 Cl. Ct. 490 (1985).
[22]   ENGBCA No. 4926, 87-2-BCA 19,833.

Other main issues that judges found exhaustive in following and understanding a critical path method are the complex networks and the relationships between activities especially if they are not well articulated or presented in the form of fragnets or sub-networks. In certain cases, the judge expressed his preference to analysis based on factual evidence, sound practical experience and common sense despite that such analysis might not be based on critical path analysis and jettisoned the approach based on flawed as-built critical path analysis.

In the case of *Great Eastern v Laing*,[23] the expert witnesses of the parties approached their analyses of the delay using two different approaches to a critical path method which attracted insightful comments from J. Wilcox due to the hypothetical manner they were presented:

> I reject Mr Celetka's evidence that the late design information either caused or contributed to the critical delay in the project. His analysis was self confessedly incomplete. He did not have the time to approach the research of this aspect of the case in the complete and systematic way, furthermore, the impacted as planned analysis delay takes no account of the actual events which occurred on the project and gives rise to an hypothetical answer when the timing of design release is compared against the original construction programme. Thus it would take no account of the fact that the design team would have been aware of significant construction delays to the original master programme, and would been able to prioritise design and construction to fit this. Furthermore, Mr Celetka in his report compares the timing of the actual design releases against an original programme which was superseded by later versions of the procurement programme on which Laing showed later dates for the provision of the information required.

In the case of *City Inn v Shepherd Construction Ltd*,[24] the defendant approach, which was a form of an as-built critical path method, was rejected by Lord Young as indicated in his judgement:

> In my opinion the pursuers clearly went too far in suggesting that an expert could only give a meaningful opinion on the basis of an as-built critical path analysis. For reasons discussed below, I am of opinion that such an approach has serious dangers of its own. I further conclude, as explained in those paragraphs, that Mr Lowe's own use of an as-built critical path analysis is flawed in a significant number of important respects. On that basis, I conclude that that approach to the issues in the present case is not helpful. The major difficulty, it seems to me, is that in the type of programme used to carry out a critical path analysis any significant error in the information that is fed into the programme is liable to invalidate the entire analysis. That seems to me to invalidate the use of an as-built critical

---

[23] Great Eastern Hotel Company Ltd v John Laing Construction Ltd. TCC (2005) All ER 368. The works were carried out by trade contractors with Laing as construction manager of the project. The dispute involved claims raised by Great Eastern in respect of project delay of about 44 calendar weeks. By way of defence, Laing made a counterclaim based upon alleged material misrepresentation and also denied culpability of the delay by blaming both other parties and other concurrent causes of delay.

[24] (2007) CSOH 190. This case concerned the construction of a hotel in Bristol under an amended JCT 80 Form. Matters in dispute included the pursuer, City Inn, seeking a declarator that the defendant was not entitled to the ontended 11 weeks time extension and even the four week extension granted by the architect.

path analysis to discover after the event where the critical path lay, at least in a case where full electronic records are not available from the contractor.

The above cases suggest that although critical path method techniques are recognised as appropriate for delay analysis, it is very important for contractors and clients or their agents to employ techniques that consider what actually happened on site based on factual evidence. Theoretical delays calculated without taking into account actual project records are unlikely to succeed. However, the cases do not seem to make things clear as to which methodology is the most acceptable by the courts.

4.12 Concurrent Delays

Concurrent delays resolution in construction disputes refers to the situation where a particular delay to the completion of the works may be attributed to two or more events. Contractors will generally have an express contractual entitlement to an extension of time for delays caused by an employer risk event such as inclement weather or instructions to vary the works, however, no such relief will be available for a contractor risk event, where the contractor itself is responsible for the delay. In construction projects it is not uncommon for a number of different events, some caused by the contractor and the others by the client or his agents, to cause delay to the works simultaneously or at different intervals.

Concurrency has been a contentious legal and technical subject in engineering and construction projects. The reason is largely due to the fact that resolving it requires the consideration of the interaction of different factors such as the time of occurrence of the delays, its length of duration and criticality, the legal principles of causation and float ownership. Its resolution also requires the consideration of the views of the parties involved and the mitigation steps taken by them such as reallocation of resources, incentives for acceleration procedures and delay-pacing strategies.

There is the absence of any readily identifiable definition, along with a failure by the courts to give any clear guidance on the most suitable or appropriate method, for considering an extension of time award when there are concurrent delays. There are quite a few methods for trying to determine an extension of time each having varying degrees of success in the courts. The most common approaches in dealing with concurrency are identified as follows:

1. The 'common sense' test where the question of causation must be treated by the application the logical principles of causation.
2. The 'but for' test. By this test, a party seeks to lay responsibility for project delay on the other party by arguing that the delay would not have occurred but for the latter actions or inactions which occurred first.
3. The 'Malmaison' approach with the view that provided one of the causes of delay in any given concurrency situation affords grounds for extension of time

under the contract, then the contractor should be given time extension not-withstanding any default on his part.

In the case of *H. Fairweather and Co. Ltd v London Borough of Wandsworth*,[25] the court, *obiter dicta*, considered that the approach of applying common sense approach to concurrency was not sufficient on its own merit and that each separate cause of delay should be assessed on its own. Another weakness of this approach is the common sense criterion relied on which could result in unfair apportionment, particularly where the competing causes are of approximate equal causative potency. Additionally, the approach may not suffice on projects that sustained multiple overlapping changes or delays with long durations because of all the assumptions that must be made regarding the remaining durations of activities being affected which means that the programme becomes too hypothetical to apply.

The 'but for' method tends to attract the most support from the contracting fraternity as it tends to support the claimant. It is based on a simple concept that the overrun would not have occurred but for the architect instruction. Although such argument are often made there appear to be few reported court cases that lends support to its use. In the case of *The Royal Brompton Hospital NHS Trust v Hammond* (2001) (*supra*), J. Seymour provided little support to the aforemen-tioned test:

> However, if Taylor Woodrow was delayed in completing the works both by matters for which it bore the contractual risk and by relevant events, within the meaning of that term in the standard form, in light of the authorise to which I have referred, it would be entitled to extensions of time by reason of the occurrence of the relevant events not withstanding its own defaults.

The 'Malmaison' approach, which is the most adopted and accepted by the industry, refers to the *Henry Boot construction (UK) Ltd v Malmaison Hotel (Manchester) Ltd* (2000) (*supra*) case. The view purported by this case is that provided one of the causes of delay in any given concurrency situation affords grounds for extension of time under the contract, then the contractor should be given time extension notwithstanding any default on his part. The approach sounds reasonable and just in the sense that denying the contractor time extension in such circumstances could make him liable to the payment of liquidated damages even though the project would have been delayed anyway due to employer default. In his judgement, J. Dyson stated:

> It is agreed that if there are two concurrent causes of delay, one of which is a relevant event, and the other is not, then the contractor is entitled to an extension of time for the period of delay caused by the relevant event notwithstanding the concurrent effect of the other event. Thus to take a simple example, if no work is possible on a site for a week not only because of exceptionally inclement weather (a relevant event), but also because the contractor has a shortage of labour (not a relevant event), and if the failure to work during that week is likely to delay the works beyond the completion date by one week, then if he

---

[25] (1987) 38 BLR 106.

considers it fair and reasonable to do so, the architect is required to grant an extension of time of one week. He cannot refuse to do so on the grounds that the delay would have occurred in any event by reason of the shortage of labour.

A slight departure from the *Malmaison* case arose in the case of *Motherwell Bridge Construction Ltd v Micafil Vakuumtechnik* (2002) (*supra*), where J. Toulmin stated, *obiter dicta*, that it is necessary to apply a test of common sense and fairness in deciding matters of extensions of time involving issues of concurrency. He considered that a full extension of time should be awarded where there is concurrent contractor caused and employer caused delay, if it is fair and reasonable to do so.

In the American approach based on US case law, the general view on concurrent delays is that the employer and the contractor are both responsible for delays to project completion, and neither party will recover financial recompense unless and to the extent that they can segregate delay associated with each competing cause somehow described this view as the easy rule and fair rule.

4.13   Concurrency and Apportionment

It is probable that the incidence of concurrent delays will increase greatly using the broad definition of concurrency as adopted by the courts, as it seems likely that most sizeable contracts will have a number of relevant events and contractor risk events, all of which have a varying effect on the completion date. In the absence of there being an identifiable dominant cause, the architect or the contract administrator will then be expected to apply the apportionment principle to the competing concurrent causes.

Where there are competing delays, and with the absence of apportionment, a claim either for an extension of time or a cross claim for liquidated damages should succeed. It anticipates for the architect or the client's agent to consider the cause or reasons for the delays and chose a dominant one, which in theory sounds perfectly reasonable. However, in practice, it often creates considerable difficulties to the extent that it could be impossible or impractical to apply when the two causes have equal potency or that the engineer cannot simply determine the difference. This rationale is based on the parties having intended that in the event of a delay one of them must be responsible.

In determining apportionment in case of concurrent delays, the following principles must be followed:

- Before any claim for an extension of time can succeed, it must be shown that the relevant event is likely to delay or has delayed the works.
- Whether the relevant event actually causes delay is an issue of fact which is to be resolved by the application of principles of common sense.
- The decision-maker can decide the question of causation by the use of whatever evidence he considers appropriate. If a dominant cause can be identified in

respect of the delay, effect will be given to that by leaving out of account any cause or causes that are not material.

- Where there are two causes operating to cause delay, neither of which is dominant, and only one of which is a relevant event, a contractor claim for an extension of time will not necessarily fail. Rather, it is for the decision-maker approaching the issue in a fair and reasonable way, to apportion the delay in completion of the works as between the relevant event and the other event.

In *John Doyle Construction v Laing Management (Scotland)*,[26] Lord Young was of the opinion that it may be possible to apportion the loss between the causes for which the employer is responsible and other causes if it can be said that events for which the employer is not responsible are the dominant cause of the loss. In such a case it is obviously necessary that the event or events for which the employer is responsible should be a material cause of the loss. Provided that condition is met, however, apportionment of loss between the different causes is possible in an appropriate case, where the causes of the loss are truly concurrent, in the sense that both operate together at the same time to produce a single consequence.

In *Doyle v Laing*, Lord Young stated:

> Apportionment in this way, on a time basis, is relatively straightforward in cases that involve only delay. Where disruption to the contractor's work is involved, matters become more complex. Apportionment will frequently be possible in such cases, according to the relative importance of the various causative events in producing the loss. Whether it is possible will clearly depend on the assessment made by the judge or arbiter, who must of course approach it on a wholly objective basis. It may be said that such an approach produces a somewhat rough and ready result. The alternative to such an approach is the strict view that, if a contractor sustains a loss caused partly by events for which the employer is responsible and partly by other events, he cannot recover anything because he cannot demonstrate that the whole of the loss is the responsibility of the employer. That would deny him a remedy even if the conduct of the employer or the architect is plainly culpable, as where an architect fails to produce instructions despite repeated requests and indications that work is being delayed. In such cases the contractor should be able to recover for part of his loss and expense, and that the practical difficulties of carrying out the exercise should prevent him from doing so.

In *City Inn v Shepherd Construction* (2007) (*supra*), the sitting trial judge was again Lord Young who adopted the approach of apportionment when it came to considering concurrent delays. He confirmed, *obiter dicta*, that if a dominant cause could not be established between two competing causes, one a relevant event and the other a contractor default, then the architect or a tribunal could apportion the effects of these delays. Lord Young views are expressed in the following statement:

> As various causes of delay are likely to interact in a complex manner, the architect must exercise his judgement to determine the extent to which completion has been delayed by relevant events. The architect must make a determination on a fair and reasonable basis.

---

[26] (2004) 1 BLR 295.

Where there is true concurrency between a relevant event and a contractor default, in the sense that both existed simultaneously, regardless of which started first, it may be appropriate to apportion responsibility for the delay between the two causes; obviously, however, the basis for such apportionment must be fair and reasonable. Precisely what is fair and reasonable is likely to turn on the exact circumstances of the particular case.

Lord Osborne, in the appeal case of *City Inn v Shepherd Construction*,[27] in agreeing with Lord Young said that where there are concurrent causes, it will be possible for an architect or other tribunal to apportion delay to the completion of the works between the competing causes, assuming that there is no evidence of a dominant cause. He further commented:

Where a situation exists in which two causes are operative, one being a relevant event and the other some event for which the contractor is to be taken to be responsible, and neither of which could be described as the dominant cause, the claim for extension of time will not necessarily fail. In such a situation, which could, as a matter of language, be described as one of concurrent causes, in a broad sense, it will be open to the decision maker, whether the architect, or other tribunal, approaching the issue in a fair and reasonable way, to apportion the delay in the completion of the works occasioned thereby as between the relevant event and the other event.

4.14   Time at Large

The time for completion of works can become at large when the contractor has been hindered or prevented by the employer from completing the works in accordance with the original contract. This results from a well-established principle of law that "no person can take advantage of the non-fulfilment of a condition the performance of which has been hindered by himself." Time being at large does not mean that the contractor has no obligation to complete the work; he has to complete in a reasonable time. Even if the employer is not entitled to liquidated damages he can still recover general damages, if he can prove that he has suffered a loss as a result of the contractor delay.

This principle is also called the prevention principle. The prevention principle can only apply where the contract does not contain a proper mechanism for allowing the employer to grant an extension of time for delay caused by the employer or anyone for whom he is responsible or if there is a mechanism and the architect does not use it. The essence of the prevention principle is that the employer cannot hold the contractor to a specified completion date if the employer has, by an act or omission, prevented the contractor from completing by that date. Instead, time becomes at large and the obligation to complete by the specified date is replaced by an implied obligation to complete within a reasonable time. It is in

---

[27]   (2010) CSIH 68 CA 101/00.

order to avoid the prevention principle that many construction contracts and subcontracts include provisions for extension of time.

Time for completion of the works can be said to be, or made, at large in the following situations:

1. No time for completion is fixed in the contract.
2. Improper administration or misapplication of the extension of time provision in the contract.
3. Waiver of time requirements.
4. The employer interference in the certification process.
5. When the extension of time provision does not confer power on the engineer and architect to extend the time for completion of the works on the occurrence of an event or events which fall within the obligation of the employer.

If the contractor incurs additional costs as a direct result of the client delay and/ or contractor delay, then the contractor should only recover monetary compensation if he is able to separate the additional costs caused by the client delay from those caused by the contractor delay. Therefore, a contractor delay should not reduce the amount of the extension of time due to the contractor as a result of the client delay and analyses should be carried out for each event separately and strictly in the sequences in which they arise.

The leading authority on the point is the decision in *Peak Construction (Liverpool) Ltd v McKinney Foundations Ltd* (1970) (*supra*), in which LJ Salmon said:

> A clause giving the employer liquidated damages at so much a week or month which elapses between the date fixed for completion and the actual date of completion is usually coupled, as in the present case, with an extension of time clause. The liquidated damages clause contemplates a failure to complete on time due to the fault of the contractor. If the failure to complete on time is due to the fault of both the employer and the contractor, in my view, the clause does not bite. I cannot see how, in the ordinary course, the employer can insist on compliance with a condition if it is partly his own fault that it cannot be fulfilled.

The time at large principle was applied in its most extreme form in *Gaymark Investments Pty Limited v Walter Construction Group Limited*.[28] In that case, the contractor was delayed in completing the work, by causes for which the employer was responsible, which constituted acts of prevention by the employer with the result that there was no date for practical completion and the contractor was then obliged to complete the work within a reasonable time. In his judgment, J. Bailey, in affirming the arbitrator decision, that Gaymark were not entitled to liquidated damages and time was set at large based on the prevention principle, said the following:

> In the circumstances of the present case, I consider that this principle presents a formidable barrier to Gaymark's claim for liquidated damages based on delays of its own

---

[28] (1999) NTSC 143.

making. I agree with the arbitrator that the contract between the parties fails to provide for a situation where Gaymark caused actual delays to Concrete Construction's achieving practical completion by the due date coupled with a failure by Concrete Constructions to comply with the notice provisions of SC19.1. In such circumstances, I do not consider that there was any manifest error of law on the face of the award or any strong evidence of any error of law in the arbitrator holding that the prevention principle barred Gaymark's claim to liquidated damages.

In *Multiplex Constructions (UK) Ltd v Honeywell Control Systems Ltd.*,[29] Honeywell argued that time had become at large and that its obligation to complete its subcontract works within a specified time had fallen away, only to be replaced by an obligation to complete the works within a reasonable time. Honeywell founded his argument on the prevention principle and argued that time had become at large mainly due to the issue of further instructions and programmes which meant that there was a delay to the finishing of the works and that Multiplex had simply failed to operate the extension of time machinery in the contract and, in the alternative, that machinery had broken down. J. Jackson, in disagreeing with this argument presented, found that "it was well established law that a party cannot insist upon the performance of an obligation, which he has prevented a promisor from performing" and the provisions of the contract clauses entitled an application for an extension of time and thus time was not at large.

4.15 Acceleration

One of the methods that contractors will employ to counteract the impact of delay and disruption is to accelerate the work on the project. This involves speeding up aspects of the project that have potential to be accelerated, either because of the nature of the task or because of the resources or capabilities of the contractor. In either case, strengths of the contractor in one area are being used to compensate for weaknesses in another area. Acceleration can also be used when there is no delay, either to counteract possible future delays that might be envisaged, or in order to achieve an early completion, but in most cases acceleration will be used to try and compensate for already experienced delays.

In construction, acceleration is not a legal term and the date against which progress is measured is usually the benchmark against which to measure the acceleration process. Acceleration may be achieved by a change in the deployment of resources, longer working hours or additional days of working with the same resources. In some cases it may be achieved by simply changing the order or sequence for carrying out the work and may therefore not cause additional cost. The phenomenon of acceleration, while beneficial for the developer, can affect a contractor's costs in a variety of ways, such as:

---

[29] (2007) BLR 195 TCC.

1. Additional labour and equipment costs arising from reduced efficiency of the expanded labour force and supplied equipment.
2. Additional delivery charges for material and equipment required at the site outside of normal work hours.
3. Costs of additional site facilities.
4. Additional costs from an advancing of the date of delivery of manufactured elements.
5. Overtime for engineers, staff and foremen.

Acceleration can also lead to benefits for the contractor, such as when regular progress of the site operations would inevitably result in liability to the developer arising from the delayed completion of the works. When given a choice to accelerate or not to accelerate, careful and deliberate evaluation of the advantages and the disadvantages of acceleration in the particular case is called for. A client may chose to reverse the impact of delays by expressly ordering to put the project back on schedule and applying the critical path method analysis or other scheduling and programming techniques, showing the cost of such a directive. In other cases, where the client resists to grant the contractor an extension of time that he is entitled to an excusable event or where an extension has been granted by the owner shorter than the contractor is entitled to, a contractor may feel compelled to accelerate the works in order to overrun the completion date set by the owner thereby avoiding exposure to liquidated damages.

Because it is generally recognised that a contractor is entitled to its full contract term to complete, costs of acceleration carried out pursuant to a direction of the client which shortens the available term are usually negotiated and agreed before the acceleration begins. Similarly, acceleration directed to be carried out to overcome delay for which the contractor is legally responsible (culpable delay) is generally accepted by the contractor when the contractor acknowledges responsibility for the delays. The contractor may not be able to recover its acceleration costs if the delays were the contractor's responsibility.

Constructive acceleration, which is the most commonly disputed form of acceleration, is said to occur when the contractor claims for an extension of time but the developer denies that request and affirmatively requires completion within the existing contract term, and it is later determined and agreed that the contractor was entitled to the extension of time. To recover under this theory the contractor must prove:

1. That an extension of time was requested for an excusable delay according to the contract provisions.
2. The client failed to grant an adequate or any extension of time.
3. That the client made it clear that completion was required within the original contract period.
4. That adequate notice had been given by the contractor to the owner advising that he was treating the owner actions as constructive acceleration.
5. And finally the proof that there had been the actual issuance of additional costs.

In *John Barker Construction Ltd v London Portman Hotel Ltd* (1996) (*supra*), one of the issues disputed was if liquidated damages could be imposed when acceleration was agreed between the parties on the sectional completion provisions of the contract. In this case delays occurred and it was apparent to all concerned that John Barker was entitled to extensions of time. After negotiations it was agreed that the work would be accelerated and John Barker would receive additional payment. J. Toulson held that the provisions of the sectional completion regarding liquidated damages were capable of continuing to have contractual force for the completion of each section, even though an acceleration agreement was in place, since it was common ground at the time of the agreement of acceleration that liquidated damages clause would still have effect.

In *Ascon Contracting Limited v Alfred McAlpine Construction Isle of Man Limited*,[30] there had been delays due to a number of causes and Ascon claimed for loss caused by acceleration measures it had undertaken. J. Hicks affirmed, *obiter dicta*, that acceleration had no precise technical meaning. Acceleration, which was not required to meet a contractor existing obligations, was likely to be the result of an instruction from the client for which he must pay. On the other hand, pressure from the client to make good delay caused by the contractor own default was unlikely to be so construed. It was held that there could be both an extension to the full extent of the employers culpable delay with damages on that basis and also damages in the form of expenses incurred by the way of mitigation, unless it was alleged and established that the attempt at mitigation, although reasonable, was wholly ineffective.

In *Motherwell Bridge Construction Limited v Micafil Vakuumtecchnik* (2002) (*supra*), the issue of acceleration was addressed in a long and complicated judgment. One of the two claims presented by Motherwell was for the acceleration costs for the work in relation to on site fabrication for hours worked in excess of normal working hours for a certain period of the project. J. Toulmin noted that Motherwell case was not that it had received any instructions to accelerate, but that it had generally been under pressure from Micafil to complete earlier and had employed additional resources to that end. There was no dispute that Micafil constantly urged Motherwell to increase its resources to meet the requested completion date. There was a term of the contract that if unexpected delays and difficulties occurred, Motherwell was required to provide additional personnel at no extra cost at the request of Micafil in order to meet the required completion date. He held, *obiter dicta*, that the delays and difficulties came within the definition of "unexpected" and Motherwell could not succeed in recovering damages for this item.

---

[30] (1999) Con LR 119. Ascon claimed that it allocated additional resources, worked for longer hours and seven days per week, and purchased and supplied duplicate plant and equipment. Ascon claimed that these acceleration measures were taken in order to mitigate the delays caused.

References

Case Law

Amalgamated Building Contractors Ltd. v Waltham Holy Cross UDC (1952) 2 All ER 452 CA
Ascon Contracting Limited v Alfred McAlpine Construction Isle of Man Limited (1999) Con LR
    119
Balfour Building Ltd. v Chestermount Properties Ltd. (1993) 62 BLR 1 QBD
Balfour Beatty Construction Limited v The Mayor and Burgess of the London Borough of
    Lambeth (2002) 1 BLR 288
British Steel Corporation v Cleveland Bridge & Engineering Co. (1981) 24BLR100
City Inn v Shepherd Construction Ltd. (2007) CSOH 190
City Inn v Shepherd Construction Ltd. (2010) CSIH 68 CA 101/00
Emson Eastern Ltd. v E.M.E. Development Ltd. (1991) 55 BLR114
Fortec Constructors v United States 8 Cl. Ct. 490 (1985)
Gaymark Investments Pty Limited v Walter Construction Group Limited (1999) NTSC 143
Great Eastern Hotel Company Ltd. v John Laing Construction Ltd. TCC (2005) All ER 368
H Fairweather and Co. Ltd. v London Borough of Wandsworth (1987) 38 BLR 106
Henry Boot Construction (UK) Ltd. v Malmaison Hotel (Manchester) Ltd. (2000) CILL 1572
    TCC
HW Neville (Sunblest Ltd.) v William Press & Son Ltd. (1981) 20 BLR 78
J. Jarvis and Sons v Westminster Corporation (1969) 1 WLR 1448 CA
John Barker Construction Limited v London Portman Hotel Limited (1996) 83 BLR 31
John Doyle Construction v Laing Management (Scotland) (2004) 1 BLR 295
John Driggs Company, Inc., ENGBCA No. 4926, 87-2-BCA 19,833
Motherwell Bridge Construction Ltd (T/A Motherwell Storage Tanks) v (1) Micafil Vakkuum-
    technik Ag (2) Micafil (2002) 8 Con LR 44
Mount Charlotte Investments Ltd. v Westbourne Building Society (1976) 1 All ER 890
Multiplex Constructions (UK) Ltd. v Honeywell Control Systems Ltd (2007) BLR 195 TCC
Natkin & Co v George A. Fuller Co 347 F. Supp. 17 (W.D. Mo. 1972)
Neodox Limited v The Borough of Swinton and Pendlebury (1958) 5 BLR 38
Peak Construction (Liverpool) Ltd v McKinney Foundations Ltd (1970) 1 BLR 111
Raineri v Miles (1980) 2 WLR 847
Tan Ah Kian v Haji Hasnan (1962) MLJ 400
The Royal Brompton Hospital v Hammond & Others (No. 7) (2001) 76 Con L.R. 148
Wells v Army & Navy Cooperative Society (1902) 86 LT 764

Books

Ansley R, Kelleher T, Lehman A (2009) Smith, currie and hancock's common sense construction
    law: a practical guide for the construction professional. Wiley, USA
Beale H, Bishop W, Furmston M (2008) Contract: cases and materials. Oxford University Press,
    Oxford
Bockrath J, Plotnick F (2010) Contracts and the legal environment for engineers and architects.
    McGraw-Hill, USA
Carnell N (2005) Causation and delay in construction disputes. Blackwell Publishing, Oxford
Delay and Disruption Protocol (2002). Society of Construction Law, UK
Eggleston B (2008) Liquidated damages and extensions of time: in construction contracts. Wiley-
    Blackwell Publishing, Oxford
Hinze J (2000) Construction contracts. McGraw-Hill, New York, USA

Pickavance K (2010) Delay and disruption in construction contracts. Sweet and Maxwell, UK
Ramsey V, Furst S (2008) Keating on building contracts. Sweet and Maxwell, UK
Reese C (2010) Hudson's building and engineering contracts. Sweet and Maxwell, UK
Wallace D (1995) Hudson's building and engineering contract. Sweet and Maxwell, UK
Wickwire J, Driscoll T, Hurlbut S, Groff M (2010) Construction scheduling: preparation, liability, and claims. Aspen Publishers, Maryland

Chapter 5
Damages and Calculation of Losses

5.1 Types of Damages and Remedies

One of the standard generalisations of the remedies of the common law of obligations is that the law of contract protects a person's expectations by putting him into a position he would have been in had those expectations been fulfilled. Since, in an exchange relationship, the parties are known to each other, the relevant expectations are those generated by a promise voluntarily given by a promisor to a promisee. If these expectations are dented by the defendant wrong, the harm suffered is treated as a variety of *status quo* loss, in which case the plaintiff is put into the position he was before the wrong was committed; as far as a money payment of damages can do this.

The court will similarly take the claimant overall position into account in determining the basis on which damages are to be assessed. It will not generally order the defendant to pay an amount which will actually make the claimant position better than it would have been if the contract has been performed. As a general rule, damages are based on loss to the claimant and not on gain to the defendant.

The modern law on the principles of damages, begins with *Hadley v Baxendale*,[1] a case in which Alderson B laid down a general rule for the parties recovery of damages in a breach of action by stating the following:

> Where two parties have made a contract which one of them has broken, the damages which the other party ought to receive in respect of such breach of contract should be such as may fairly and reasonably be considered as either arising naturally, that is according to

---

[1] (1854) 9 Exch 341. A shaft in Hadley's mill broke rendering the mill inoperable. Hadley hired Baxendale to transport the broken mill shaft to an engineer in Greenwich so that he could make a duplicate. Hadley told Baxendale that the shaft must be sent immediately and Baxendale promised to deliver it the next day. Baxendale did not know that the mill would be inoperable until the new shaft arrived. Baxendale was negligent and did not transport the shaft as promised, causing the mill to remain shut down for an additional 5 days.

A. D. Haidar, *Global Claims in Construction*,
DOI: 10.1007/978-0-85729-730-3_5, © Springer-Verlag London Limited 2011

the usual course of things, from such a breach of contract itself, or such as may be reasonably supposed to have been in the contemplation of both parties, at the time they made the contract, as the probable result of the breach of it.

This case was a useful devise for controlling jury verdicts on damages. As there was now a test for damages, the question became one of law and not fact, and was reviewable on appeal. This allowed the damages to a contract doctrine to develop.

The action for damages is always available, as of right, when a contract has been broken. It should, from this point of view, be contrasted for specific relief and for restitution, which are either subject to the discretion of the court or only available if certain conditions are satisfied. In *Moschi v Lep Air Services Ltd*,[2] Lord Diplock stated that "in regards to when the basic principle which the law of contract seeks to enforce is that a person who makes a promise to another ought to keep his promise" and that the basic principle is subject to remedy for the failure by a promisor to perform his promise by entering into a contract with him which creates an obligation to perform it.

In that regard the doctrine for damages was identified by Lord Diplock by stating:

If he does not do so voluntarily there are two kinds of remedies which the court can grant to the promisee. It can compel the obligor to pay to the obligee a sum of money to compensate him for the loss that he has sustained as a result of the obligee's failure to perform his obligation. This is the remedy at common law in damages for breach of contract.

Loss includes any harm to the person or property of the claimant, and any other injury to his economic position. Harm to property covers damage to or destruction of particular things; while injury to the claimant economic position includes any amount by which he is worse off that he would have been if the contract had been performed. For example, if a contractor in breach of contract fails to finish his works, or to deliver them on time, the client *prima facie* suffers loss in not having the works done, or in not having them at the agreed time. This falls under the doctrine of what constitutes loss.

Philips v Ward[3] is a case where a surveyor in breach of contract failed to draw his client attention to the fact that the roof timbers of a house, which the latter was about to buy, were rotten. It was held that the client was not entitled to damages based on the cost of making the defects good. Such an award would put him into a better position than that in which he would have been if the contract had not been broken.[4] Hence the client was entitled to recover only the difference between the price that he paid and the value of the house when he bought it.

---

[2] (1972) 2 All ER 393. Where A undertook that the principal debtor B would carry out his contract so that if B failed to act as required by his contract, he not only broke his own contract but also put the guarantor A in breach of his contract of guarantee.

[3] (1956) 1 W.L.R471.

[4] "for it would enable to have a new roof with new timbers, which would be less expensive to maintain than an old roof with sound timbers."

In *Teacher v Caddell*,[5] a financier broke a contract to invest £15,000 in the business of a timber merchant and instead invested in a distillery. It was held that the timber merchant damages were based on the loss to his business and not on the much larger profits which the financier had derived from the distillery. Similarly, in *The Siboen v The Sibotre*,[6] where a ship owner in breach of contract withdraws his ship from the charter party, damages are based on the charterer loss, and not on any profit that the ship owner may make from other employment of the ship.

Punitive or exemplary damages can be awarded in certain cases. The purpose of such damages is to express the court's disapproval of the defendant conduct. As general, punitive damages cannot be awarded in a purely contractual action since the object of such an action is not to punish the defendant but to compensate the claimant. Punitive damages are not available even though the breach was committed deliberately and with a view to profit. If the court is particularly outraged by the defendant conduct it can sometimes award damages for injury to the claimant feelings.

In *Whiten v Pilot Insurance Co.*,[7] the court decided, *obiter dicta*, the following:

- The handling of claims in a bad faith manner should be subject to punitive or aggravated damages.
- Punitive damages can be awarded for an actionable wrong that need not be tort but be a breach of a good faith duty.
- Punitive damages are not to be ordered in the normal course of things; there must be exceptional circumstances.

The doctrine of reliance loss is to put the claimant into the position in which he would have been if the contract had never been made, by compensating him for expenses incurred or other loss suffered in reliance on the contract. Recovery of damages on the basis of reliance loss is not available if the defendant can prove that the plaintiff had entered into a losing contract and would not have been able to recoup the expenditure even if the defendant had performed all its obligations.

In *McRae v Commonwealth Disposals Commission*,[8] the defendants were held liable for breach of a contract that there was a wrecked tanker lying in a specified position; and the claimants recovered inter alias, the £3,000 which it had cost them to send out a salvage expedition to look for the tanker. It was held by Lord Dixon that as a general rule damages may be recoverable on the basis of reliance loss

---

[5] (1889) 1F (HL) 39.

[6] Occidental Worldwide v Skibs A/S Avanti (the Siboen and the Sibotre) (1976) 1 Lloyd's Rep. 293.

[7] (2002) S.C.R. 595. Plaintiff's house caught fire. Pilot Insurance Co. aggressively pursued a fraud investigation and pursued a trial on trumped up arson allegations. The jury award of $1 m for punitive damages was restored.

[8] (1951) 84 C.L.R. McRae wasted money searching for wreck. His claim for the loss of profits expected from a successful salvage was dismissed as too speculative, however, reliance damages were awarded for wasted expenses.

where there is no way of quantifying the expectation loss; or no profit will be made on the contract.

5.2 Liquidated Damages

Construction contracts commonly make express provision for the amount of damages which are to be paid by the contractor in the event of late completion. The deduction of fixed pre-agreed amounts based on non-compliance with key performance indicators or criteria is also a common method for providing incentive for a contractor or supplier to perform well and on time. This acts as a protection for the contractor against unliquidated general damages claims and enables the contractor to fix the price for risk.

In addition, liquidated damages provisions avoid the difficulty, time and expense involved in proving and assessing the actual loss which a party will suffer in the event of a breach, either by delay in completion or by failure to comply with performance criteria, and thus enable the parties to know in advance the potential financial consequences of such a breach.

As a general rule, liquidated damages clauses must be constructed strictly *contra proferentem*. Therefore, where the clause is ambiguous and all other methods of construction have failed to resolve the ambiguity, the court may construe the words against the party seeking to rely on the clause. There are a number of grounds for which a liquidated damages clause may be held to be unenforceable, including ambiguity, their penal nature and failure to provide for an extension of time for acts of prevention by the employer and failure to comply with contractual procedures.

The effect of attacking the validity of a liquidated damages clause is, depending on the wording of the contract and perhaps the reason for the unenforceability of the clause, to remove the readymade and pre-set damages mechanism and substitute it with a liability to pay general or unliquidated damages, which could, subject to proof and ascertainment, be significantly higher than the liquidated rate. If, however, there is an exclusive remedy clause in the contract, the employer may have no right to recover whatsoever if the contractor is in culpable delay and the liquidated damages clauses is unenforceable.

The question of whether a liquidated damages clause is an exhaustive remedy for the employer is technically a question of construction of the relevant clauses in the circumstances of the case but it will be exceptional for it not to be an exclusive remedy for the breach to which it applies. When the contractor delay also constitutes a breach of other contractual provisions, then it may be possible to claim unliquidated damages for those breaches. Therefore, where a contractor does not proceed diligently with the works and the liquidated damages clause is drafted in such a way so as to relate only to damages for late completion, then the contractor may be exposed to liability for damages for failure to proceed diligently with the works, in addition to the liquidated damages.

The following passage of LJ Salmon in *Peak Construction (Liverpool) Ltd v McKinney Foundation Ltd.*[9] is instructive in regard to the definition of liquidated damages:

> The liquidated damages clause contemplates a failure to complete on time due to the fault of the contractor. It is inserted by the employer for his own protection; for it enables him to recover a fixed sum as compensation for delay. No doubt if the extension of time clause provided for a postponement of the completion date on account of delay caused by some breach or fault of on the part of the employer, the position would be different. This would mean that the parties had intended that the employer could recover liquidated damages notwithstanding that he was partly to blame for the failure to achieve the completion date, and the contractor would be liable to pay liquidated damages for delay as from the extended date.

In *Bramall and Ogden v Sheffield City Council*,[10] the council during the course of the project decided to take possession of the houses as they were completed with the architect issuing practical completion certificates although the contract did not allow for this since the parties had not agreed or entered into a sectional completion supplement. In presenting his case, the contractor argued, *inter alia*, that there was no express provision for sectional completion in the contract and that by taking over possession of each house as it was completed, the employer rendered the liquidated damages clause inoperable and the employer deduction of liquidated damages was unlawful. The court confirmed with this stated argument and held that there was a discrepancy in calculating liquidated and ascertained damages. Since they were expressed per dwelling this could not be reconciled with partial possession; liquidated and ascertained damages must be applied in accordance with the whole of the works. However the court explained, *obiter dicta*, this could be applicable if the council had opted for not deducting all the damages but decided to *pro-rata* the damages to the works completed.

In *Regional Construction v Chung Syn Kheng*,[11] the nominated electrical sub-contractors caused a delay to the main contract works, as a result of which the employer deducted money from the sum due to the main contractor. According to Sir Roberts CJ:

> This means that the amount provided for liquidated damages will only be enforced in favour of the plaintiff if it can be shown that this amount was a genuine pre-estimate of the damages likely to flow from the specified breach. The amount of loss or damage which has actually occurred must be a major factor in deciding whether the amount provided for was an honest pre-estimate of the likely loss or damage. If the actual loss or damage suffered is very much less than the sum agreed, the court will refuse to enforce the agreement to pay a specified sum by way of liquidated damages.

---

[9] (1970) 1 BLR 111.

[10] (1985) 1 Con LR 30. Within the contract the council had inserted for liquidated and ascertained damages for late completion at the rate of £20 per week for each complete house.

[11] Regional Construction Sdn. Bhd. v Chung Syn Kheng Electrical Co. Bhd. (1987) 2 MLJ 763.

In *Bovis Construction v Whatlings Construction Ltd*,[12] the issue before the court concerned the limitation of liquidated damages and the determination of the contract due to the contractor's failure to proceed regularly and diligently. Lord Hope emphasised, *obiter dicta*, that it is not only contract provisions fixing a level of damages which may be treated in law as penalties. The same principle applies to clauses under which, in the event of breach, the contractor is to forfeit tools and materials, all money due under the contract or all retention money. The court held, *obiter dicta*, that such provisions are penalties, and therefore un-enforceable, where it is clear that the amount to be forfeited cannot represent a genuine pre-estimate of the client's likely loss. Liquidated damages provisions must be construed strictly so that any ambiguity will be construed *contra proferentem*, against the interest of the party seeking to rely upon the clause.

5.3 Liquidated Damages: An American Case Law Approach

It has been accepted law in the United States for many years that in order to recover for liquidated damages for delay a party must prove three elements:

(1) Liability of the party against whom it is making the claim.
(2) Causation as of facts.
(3) Damages being liquidated or unliquidated damages are within the terms of the contract.

Proof of liability and damage alone does not entitle a party to recovery. Causation, however, can be a thorny issue as multiple causes of delay pose particular problems for contractors seeking to avoid exposure to liquidated damages and to obtain compensation for delay and disruption. The court will examine the facts of the case under both the rule that forbids apportionment and the rule that permits apportionment of liquidated damages.

American law treats these two circumstances differently. In the case of concurrent delay, a contractor will generally receive what would be summarised as '*time, but no money*'. The more complex issue arises in relation to sequential delays, where one party and then the other cause different delays *seriatim* or intermittently, the contractor then tries to rely on the '*rule against apportionment*' which is derived from *United States v United Engineering and Constructing Co.*[13] and is to the effect that "where delays are caused by both parties to the contract the court will not attempt to apportion them, but will simply hold that the provisions of the contract with reference to liquidated damages will be annulled."

---

[12] Bovis Construction (Scotland) Ltd v Whatlings Construction Ltd (1995) 49 Con LR 12.

[13] 234 U.S. 236 (1914). In this case the contractor accepted a reduced payment under protest. He was able to recover liquidated damages withheld by the government after showing that much of the delay caused was by other contractors performing on the same contract.

In this case, Mr. Justice Day relied on English law to provide his judgment:

Where the original contract for government work provided for liquidated damages for delay beyond a specified date but supplemental contracts contained no fixed rule for the time of completion, the government is limited in its recovery to the actual damages sustained by reason of the delay for which the contractor was responsible. It is the English rule, as well as the rule in some of the states, that where both parties are responsible for delays beyond the fixed date, the obligation for liquidated damages is annulled, and unless there was a provision substituting a new date, the recovery for subsequent delay is limited to the actual loss sustained. Where the government has, by its own fault, prevented performance of the contract, and thereby waived the stipulation as to liquidated damages, it cannot insist upon it as a rule of damages because it may be impracticable to prove actual damages.

In *Robinson, Administrator of Robinson v The United States*,[14] the issue in this case arose out of stipulations in construction contracts obliging the contractor to pay liquidated damages for delay. A public building contract obliged the contractor to pay liquidated damages for delays on a daily basis not caused by the government. Delays were attributable to both parties. It was held that the government were entitled to damages for the part of the delay specifically found by the Lower Court to have been due wholly to the fault of the contractor. Mr. Justice Brandeis said that:

Stipulations in construction contracts obliging the contractor to pay liquidated damages for each day's delay are appropriate means of inducing due performance and of affording compensation in case of failure to perform, and are to be given effect according to their terms. The construction of the contract and the findings of fact are clear. If the provision for liquidated damages is not to govern, it must be either because, as matter of public policy, courts will not, under the circumstances, give it or because, in spite of the explicit finding, no day's delay can, as matter of law, be chargeable to the contractor where the government has caused some delay. Neither position is tenable.

In *Sun Shipbuilding and Drydock Company*,[15] the opinion delivered by Mr. Kennedy was clear when he emphasised, *obiter dicta*, that, for purposes of assessing liquidated damages, if an excusable cause of delay in fact occurs, and if that event in fact delays the progress of the works as a whole, the contractor is to be liable for damages after the entitlement to an extension of time commensurate with the delay, notwithstanding that the progress of the work was concurrently slowed down by want of diligence, lack of proper planning or some other inexcusable omission on the part of the contractor. The contractor, in this instance, is said to be excused of liability for liquidated damages for the period of the extension which otherwise would have been payable to the client.

---

[14]   (1922) 261 US 486.

[15]   ASBCA 11300, 68-1 B.C.A (1968). This case relates mainly to disruptions due to inclement weather during the course of the project.

The issue was addressed in detail in *R P Wallace Inc. v The United States*[16] where the contractor alleged that some of the delays were the fault of the employer and sought thereby to avoid liquidated damages. The court came out strongly in favour of allowing liquidated damages even though the Navy was partially at fault. However, the court reduced the number of late penalty days from 250 to 229 by apportioning responsibility for the delay between the Navy and the contractor.

5.4 Penalties

It is a well-accepted principle that liquidated damages provisions must, in both their amount and operation, constitute a genuine pre-estimate of the loss which the employer is likely to suffer. The clause will be invalid and unenforceable if the amount or its operation is a penalty and, therefore, the court will only enforce the sum identified where it represents a proper assessment of the loss.

The distinction between liquidated damages and a penalty is a matter of construction for the court, and it is for the person from whom the damages are claimed to show that the clause is in fact a penalty. A clause that is found to be penal is generally invalid, and it is an unusual feature of the law of contract that the court will strike down penalty clauses, while usually permitting other clauses which have been freely agreed between the parties even if those clauses are unduly harsh.

The use of the words penalty or liquidated damages in the contract is not conclusive or determinative of whether it is in fact one or the other. The court must find out whether the payment stipulated is in truth a penalty or liquidated damages. In other words, the parties attempt to call it liquidated damages is not conclusive and the court can still determine that it is a penalty. The provisions to consider when addressing penalties as to liquidated damages in construction contracts can be summarised by the following two criteria:

1. The essence of a penalty is a payment of money stipulated as against the offending party; the essence of liquidated damages is a genuine covenanted estimate of damages to the innocent party. The key is to attempt to estimate what actual loss would be caused by the breach by the offending party and further, whether it is fair and reasonable.
2. The question whether a sum stipulated is penalty or liquidated damages is a question of construction to be decided upon the terms and inherent circumstances of each particular contract, judged at the time of the making of the contract, not at the time of the breach.

---

[16] (2004) 63 Fed Cl 402. Plaintiff contractor, a construction firm, contracted with defendant United States acting through the Department of the Navy for renovation and repair work to a New Orleans naval facility. Because the contract was not completed on time, the Navy assessed liquidated damages. The contractor filed suit for remission of those liquidated damages, as well as compensable damages, due to government-caused delay and disruption.

To assist this task of construction of such provisions various tests have been suggested:

1. It will be held to be penalty if the sum stipulated for is extravagant and unconscionable. An applicable measurement test would show that this loss is greater than any loss that could have occurred under the same circumstances.
2. It will be held to be a penalty if the breach consists only in paying a sum of money, and the sum is greater than the sum which ought to have been reasonably paid (i.e., greater than the actual losses of the innocent party).
3. There is a presumption that it is penalty when a single sum is payable by way of compensation to different causes when some have no significant influence on the damages that have occurred.
4. The time for assessment or construction of the provision as either a genuine pre-estimate of damages or a penalty is as at the time of the contract. In other words, if it is a genuine re-estimate of damages or losses of the owner at the time the contract is entered, then it will likely be valid and enforceable.

The case of *Commissioner of Public Works v Hills*,[17] concerned a construction contract which provided that the contractor would forfeit retention monies as and for liquidated damages for late completion. It was held that this could not be considered to be a genuine pre-estimate of loss as the sum would increase with no relation to the cost of completion. Indeed, retention monies ordinarily increase as the cost of completing the works decreases. Lord Dunedin formulated the test for penalty clauses as follows:

> The general principle to be deduced is that the criterion of whether a sum, be it called penalty or damages, is truly liquidated damages, and as such not to be interfered with by the court, or is truly a penalty which covers the damage if proved, but does not assess it, is to be found in whether the sum stipulated for can or cannot be regarded as a 'genuine pre-estimate' of the creditor's probable or possible interest in the due performance of the principal obligation.

The various tests of causation to assist in identifying a penalty as opposed to damages have been established again by Lord Dunedin in *Dunlop Pneumatic Tyre Co. v New Garage and Motor Co. Ltd*.[18] Lord Dunedin reached the conclusion that if the sum to be paid for a breach of the contract was substantial and arbitrary and bore no relation to the potential loss of the other party, it was, therefore, a penalty. He stated:

> The essence of a penalty is a payment of money stipulated as in terrorem of the offending party; the essence of liquidated damages is a genuine covenanted pre-estimate of damage. The question whether a sum stipulated is penalty or liquidated damages is a question of construction to be decided upon the terms and inherent circumstances of each particular contract, judged of as at the time of the making of the contract.

---

[17] (1906) AC 368.
[18] (1915) AC 79.

In *Dunlop Pneumatic Tyre Co v New Garage and Motor Co Ltd,* Lord Dunedin identified the principles to be applied in order to establish a penalty as:

(a) It will be held to be penalty if the sum stipulated for is extravagant and unconscionable in amount in comparison with the greatest loss that could conceivably be proved to have followed from the breach. (b) It will be held to be a penalty if the breach consists only in not paying a sum of money, and the sum stipulated is a sum greater than the sum which ought to have been paid. (c) There is a presumption that it is penalty when a single lump sum is made payable by way of compensation, on the occurrence of one or more or all of several events, some of which may occasion serious and others but trifling damage. (d) It is no obstacle to the sum stipulated being a genuine pre-estimate of damage, that the consequences of the breach are such as to make precise pre-estimation almost an impossibility. On the contrary, that is just the situation when it is probable that pre-estimated damage was the true bargain between the parties.

In the *Robophone Facilities Ltd v Blank,*[19] this prohibition of the penalty clause is said to be consistent with the law treatment of damages as damages are compensatory and that to allow a clause which allows recovery of damages in excess of the actual loss suffered or sufferable would be wrong. The damages would be more of a deterrent, designed to discourage breaches of contract or to secure performance by the contractor, than compensation. LJ Diplock stated:

The onus of showing that such a stipulation is a 'penalty clause' lies upon the party who is sued upon it. The terms of the clause may themselves be sufficient to give rise to the inference that it is not a genuine estimate of damage likely to be suffered but is a penalty. Thus it may seem at first sight that the stipulated sum is extravagantly greater than any loss which is liable to result from the breach in the ordinary course of things. This would give rise to the prima facie inference that the stipulated sum was a penalty. But the plaintiff may be able to show that owing to special circumstances outside 'the ordinary course of things' a breach in those special circumstances would be liable to cause him a greater loss of which the stipulated sum does represent a genuine estimate.

In *Philips Hong Kong v A-G of Hong Kong,*[20] Lord Woolf was of the opinion to the effect that the courts should not try too hard to find a penalty clause and, for example, the identification of circumstances where the liquidated damages would amount to an over-recovery by the plaintiff was not necessary. The key question remains whether the liquidated damages were a genuine pre-estimate of loss at the time they were set. He stated the following:

Except possibly in the case of situations where one of the parties to the contract is able to dominate the other as to the choice of the terms of a contract, it will normally be insufficient to establish that a provision is objectionably penal to identify situations where the application of the provision could result in a larger sum being recovered by the injured party than his actual loss. Even in such situations so long as the sum payable in the event of non-compliance with the contract is not extravagant, having regard to the range of

---

[19] (1966) 1 W.L.R. 14.

[20] (1993) 61 BLR 41. The contract works included the design, supply, testing, delivery, installation and commissioning of a processor-based supervisory system for the approach roads and twin tube tunnels which were to be constructed under Smuggler's Ridge and Needle Hill Mountains in the New Territories as part of the Shing Mun Section of the project.

losses that it could reasonably be anticipated it would have to cover at the time the contract was made, it can still be a genuine pre-estimate of the loss that would be suffered and so a perfectly valid liquidated damage provision. The use in argument of unlikely illustrations should therefore not assist a party to defeat a provision as to liquidated damages.

5.5   Quantum Meruit and Restitution

The principle of restitution where a claim under this principle is not strictly one for damages since its purpose is not to compensate the claimant for a loss, but to deprive the defendant of a benefit. A restitution claim differs from a loss of bargain claim, which is meant to put the claimant into the position in which he would have been if the contract had been performed. It also differs from a claim for reliance loss, which is meant to put the claimant into the position in which he would have been if the contract had not been made, and which will often leave the defendant in a worse position. A claim for restitution in construction contracts is usually pursued under the doctrine of *quantum meruit*.

In essence, *quantum meruit* claims flow from the principle of unjust enrichment. For there to have been unjust enrichment, three things must be established; firstly, the principal must have been enriched by the receipt of a benefit; secondly, that benefit must have been gained at the contractor's expense, and, thirdly, it would be unjust in the circumstances to allow the principal to retain the benefit. Here since, *quantum meruit* is an action for payment of the reasonable value of services performed. It is used in various circumstances where the court awards a money payment that is not determined, subject to what is said below, by reference to a contract. Put in the converse, a claim on a *quantum meruit* cannot arise if there is an existing contract between the parties to pay an agreed sum.

In a *quantum meruit* claim the court awards a money payment for works performed in the absence of a contract or when there has been a contract but it has been frustrated, made void, terminated or is otherwise unenforceable. That is not to say, however, that a *quantum meruit* claim cannot be made where there is a contract on foot. In this regard, *quantum meruit* claims can be made where:

- There is a contract but no price is fixed by that contract. Where a contractor does work under an express or implied contract, and the contract fixes no price or pricing mechanism for that work, the contractor is entitled to be paid a reasonable sum for his labour and for the materials he has supplied.
- Quasi-contract. This may arise where, for example, a contractor agrees to start work on-site while still negotiating with the principal as to, at least, the essential terms of the contract. Those negotiations subsequently fail. Generally, the cases support the proposition that, in such circumstances, the principal has an obligation to pay a reasonable sum for the work done. Predictably, however, it does not apply to all cases of failed contracts.
- Works have been carried out under a void, unenforceable or terminated contract. Where a contract has been repudiated by the client, a contractor must elect to

accept the repudiation and terminate the contract before commencing proceedings.

It is one thing to establish that a contractor may be entitled to make a *quantum meruit* claim; it is quite another to determine the extent of the entitlement. The basic reason for this is that courts have not provided clear guidelines to assist in determining what is a reasonable sum; although it is clear that the contractor ought to be paid a fair commercial rate for the work done in all the relevant circumstances. However, the essential problem remains the lack of precision in deciding what is reasonable in the circumstances. What might be reasonable for the client is to be unlikely for the contractor and vice versa. The contractor, on the other hand, may want the reasonable sum to reflect the value of the work to the client. At other times, the contractor might take the view that actual costs, plus an allowance for profit and overheads, should form the basis of a reasonable sum while the principal might adopt the view that what is reasonable should be determined by reference to the failed contract.

In *William Lacey v Davis*,[21] the contractor submitted a tender for the rebuilding of a damaged premise. The tender was not accepted but, in the belief that the contract would be placed with him, the contractor subsequently prepared various further estimates and schedules which the client made use of. The contractor even ordered some materials. Ultimately, the client did not place a contract with the contractor and instead sold the property. The contractor sued and the court found that there was no contract in place. However, the court decided that the contractor was entitled to a reasonable sum for the work carried out on a *quantum meruit* basis.

This principle was applied in the Australian case of *Sabemo Pty Ltd.v North Sydney Municipal Mutual Council*[22] In this case, a local council had invited submissions from developers for the re-development of land. The plaintiff's scheme was chosen but no formal contract was entered into. It was subsequently found by the court that there was no contract between the parties. However, the plaintiff developer had carried out a considerable amount of work in developing plans and negotiating with relevant authorities. Ultimately, the council decided not to pursue the scheme and the developer was sued for the work done. J. Sheppard said the following:

> Where two parties proceed upon the joint assumption that a contract will be entered into between them and one does work beneficial for the project and thus the interests of the two parties, which work he would not be expected, in other circumstances, to do gratuitously, he will be entitled to compensation or restitution if the other party unilaterally decides to abandon the project, not for any reason associated with bona fide disagreement concerning the terms of the contract to be entered into, but for reasons which, however valid, pertain only to his own position and do not relate at all to that other party.

J. Goff in *British Steel Corporation v Cleveland Bridge & Engineering Co.*[23] had to consider the situation of *quantum meruit* and restitution when a

---

[21] William Lacey (Hounslow) v. Davis (1957) 1 W.L.R. 932.

[22] (1977) 2 N.S.W.L.R. 880.

[23] (1981) 24 BLR 100.

quasi-contract was deduced by the parties based on the issuance of a letter of intent. He said this:

> In most cases where work is done pursuant to a request contained in a letter of intent, it will not matter whether a contract did or did not come into existence; because if the party who has acted on the request is simply claiming payment, his claim will usually be based upon a quantum meruit, and it will make no difference whether the claim is contractual or quasi-contractual. Of course, a quantum meruit claim (like the old actions for money not received and for money paid) straddles the boundaries of what we now call contract and restitution; so the mere framing of a claim as a quantum meruit claim, or a claim for a reasonable sum, does not assist in classifying the claim as contractual or quasi-contractual.

The case law shows that the extent to which *quantum meruit* may be allowed notwithstanding the existence of a contract remains somewhat fluid. In *Pavey and Matthews Pty Ltd v Paul*,[24] the High Court held, by a majority, that Pavey's action on a *quantum meruit* rested on a claim for restitution, or a claim based on unjust enrichment, arising from one party acceptance of benefits accruing from the other party's performance. From this case, it was held, *obiter dicta*, that a contractor is entitled to *quantum meruit* provided four conditions are satisfied, namely that:

1. There is no existing enforceable contract governing payment.
2. The principal has been enriched by the receipt of some benefit without payment.
3. The principal enrichment has occurred at the expense of the contractor.
4. It is unjust for the principal to retain the benefit without recompensing the contractor.

5.6 Claim Calculations: Direct and Indirect Losses

In a delay and disruption claim many items are considered as losses depending on the construction job undertaken, the contract in place and the types of losses incurred. The main items that a contractor usually relates his losses to are his direct and indirect costs which are called site overhead and off-site overhead costs, respectively. These costs are directly affected by the delays and the disruptions experienced on-site and usually contractors can successfully recover these costs in case of relevant events in a non-culpable delay periods or if they can prove disruption. Other costs, that are incurred by contractors can be claimed but are more difficult to prove and for the court to approve, relate to materials, subcontractors claims, labour, equipment, general financial costs and consequential losses.

Within the administrative structure of many contracting organisations, there is no clearly defined separation between direct costs and indirect costs. If the service or cost would have been incurred in the operation or maintenance of the

---

[24] (1987) 69 ALR 577.

contractor's organisation, irrespective of any specific contract the contractor may have been awarded, then it is an indirect cost. If, on the other hand, the service was required to be performed as part of the work done by the contractor under a specific contract, then it is a direct cost and should be included in the total cost of the execution of that contract. For example, contract managers, estimators, marketing personnel, safety and industrial officers are normally off-site overheads, although their time may occasionally be charged to a particular project for specific services. Off-site overhead expenses also include the general administrative and indirect costs of a contractor not attributable to a particular contract. These can include design work performed by the contractor or a design firm appointed by the contractor, procurement and purchasing, administrative works in relation to leasing and insurance, off-site maintenance and accounting.

A contractor overheads are normally covered by the income of the business as a whole and, where the completion of one contract is delayed, the contractor may claim to have suffered a loss arising from the diminution of the income from the contract and hence, the turnover of the business. The claim would include therefore that, due the delay and disruption caused by the client and his agents, the contractor's workforce would have had the opportunity of being employed on another contract, with the result that it would have contributed towards the costs during the overrun period. Also, if the contractor can demonstrate that engineers, foremen and other staff, who would otherwise have been gainfully employed on other sites, had to devote time to dealing with the disruption or delay, he may proceed a claim for that instance as well.

It is often asserted that when a contractor has several contracts in progress simultaneously, each contract contributes a proportionate amount to his total off-site overhead based on the individual contract amounts and the durations of each contract. Therefore, when one contract is suspended or extended due to a delay, the overall off-site overhead continues to be incurred. It is usually not increased or decreased as a consequence of the delay. It is often argued that the contribution towards the overheads anticipated to be earned from that particular contract is reduced. Notwithstanding that the contract is delayed, the amount of money earned by the contract is ultimately the same, even greater, and hence the contribution towards the total off-site overhead cost will not be less. The result is that there is no loss of recovery for off-site overhead costs which can be attributed directly to the delay. Exactly the same logic is often used by contractors to claim alleged loss of profits which, in most cases, is rejected by the client and the courts. Provided a contractor recovers the direct cost of the delay, the profit earned from a contract will not be diminished by the delay.

The accurate assessment of head office overheads, which is part of the indirect costs, on any particular contract claim is very difficult. It is highly dependent on the structure of the particular organisation, the corporate overhead of the organisation, the degree of authority given by senior management to the site personnel, the nature of the delay and the nature of the direct cost of the delay. If there are direct costs, including payment of workers, extended hire of plant and increased insurance, there will usually be some additional off-site overheads. These may include additional computer time for the processing of payment claims, extra hours for head office clerical staff, an extended period of bank overdraft and even costs for extra site visits by the managing director. The salary of the managing director will not be affected by the delay, but transport costs to pay extra site visits could be an extra cost.

5.7 Proof of Losses

Under a loss and expense claim, a contractor must prove actual costs incurred for overheads; a contractor may be able to recover extra costs by way of common law if the contractor can prove that, if the delay had not occurred, the contractor would not have suffered some loss that is not included in direct cost. There are difficulties in establishing such a claim, including principles of mitigation and foreseeability. The contractor most likely to succeed is the very small contractor who, because of the delay and disruption, is prevented from undertaking other works and hence incurs a loss of income. Larger contractors will usually be able to recruit other staff or subcontractors to perform other projects and the delays and disruptions have effect on the project itself and would not affect the contractor global operation.

Subcontractor's site overhead costs are generally less than those of a head contractor, which is generally responsible for most site services and facilities. Loss and expense claim from subcontractors are likely when the delayed subcontract was a substantial part of the subcontractor earnings for that time period and the subcontractor was prevented by the delay from undertaking other profitable work during the delay.

As for any claim, a contractor has to establish that reasonable steps have been taken to mitigate losses. Subcontractor claims can often be avoided by the subcontractor undertaking other works. This is the nature of the operations of many subcontractors in that they have many clients at one period of time and proper planning by them can both significantly mitigate losses and maintain their projected turnover to recoup what head office overheads they may have. Whether a subcontractor is nominated or selected by the client does not change the fact that there is a contractual arrangement direct between the contractor and the subcontractor. The same principles apply as for any subcontract claim.

Labour and equipment are two distinguished parts of a claim providing disruptions occur. These categories do not apply in a delay claim as it is assumed that the same resources qualitatively and quantitatively are consumed to achieve the scope of work even though the duration is longer. As when disruption occurs many items of equipment and labour can be frozen for a certain period of a project or reallocated within the same site or other construction sites to mitigate the effects of the disruption.

In certain claims, claimants can allow for consequential or indirect losses which are damages that parties may have reasonably contemplated at the time they made the contract as a probable result of a breach. (While direct losses flow naturally from a breach of contract, consequential losses are a step removed from the transaction and its immediate effects, rather than being a direct consequence of a breach). These costs include loss of profit, financial costs and losses in general that cannot be categorised but incurred by the contractor. Some courts have held that a party relying on a consequential loss clause must have actual knowledge at the time of the contract of special circumstances likely to cause the loss; otherwise the damages are too remote to be recoverable. It is important to ensure that consequential loss clauses are properly drafted to protect the party interests, including consideration of the following:

- The types of losses to be excluded, which should generally be specifically and clearly defined.
- Third party indemnities, third party liabilities, insurance indemnities, criminal acts, fraud, willful default and gross negligence.
- The effect on liquidated and general damages.
- Statutory prohibitions on exclusions.
- The effect on insurance coverage.

Loss of profits may constitute an indirect loss where these losses are outside the ordinary course of things and a party has actual knowledge of these potential losses. The manner in which a client has breached a contract will affect the complexity of the contractor claim for lost profits. The owner or developer may have failed to proceed with a project, or may cancel a project mid-construction, depriving contractors of the profits from the work they were contracted to do but which is not proceeded with. Alternatively, and more frequently, the client may complicate or delay the work of a contractor, reducing the profit of the contractor. In the case of a delayed project, the complexity of a claim for lost profit will often be greater than for a project which has never been started.

5.8   Claim Loss Components

In general a contractor can claim for delays and disruptions in the following categories:

Direct Costs. These costs are sometimes known as preliminaries costs, but are most often referred to as site overhead costs. They must be distinguished from the direct costs of construction activities. Each construction project entails certain indirect expenses that are charged directly to the job. These include typically costs of supervision, rental and depreciation of site offices, plant and machinery, site services, insurance and bond premiums.

When these costs are incurred as a result of delay and disruption for which a contractor is entitled to reimbursement, the contractor is usually entitled to recover an amount to cover these costs. To minimise the administrative effort required to provide actual cost information in support of the direct costs in a delay and disruption claim, a contractor will often provide estimated daily or weekly costs for the staff, plant and facilities involved. If good site records exist, it is usually possible to check such claims in broad terms and highlight inconsistencies in a contractor claims. The onus is on the contractor to prove the amount of such claims. An examination by both parties of cost records may become essential.

Site direct costs cover the cost of the items a contractor must provide during construction, including:

1. Salaries of site supervisory staff, including accommodation and travelling expenses.
2. Labour engaged on a part-time basis on these activities.
3. Skilled labour wages and overtime expenses.
4. General construction plant.
5. Major temporary works such as dewatering equipment, water supply, shoring and underpinning.
6. Small tools and consumables.
7. Site supply services including power, water, air and telephones.
8. Site offices, amenities, workshops and stores.
9. Site office expenses including couriers, postage, telephones and copying.
10. Insurances, security charges and long service levy charges.

Indirect Costs. Costs claimed as indirect costs may in reality be contract-related services which are performed away from the site. These works relate to procurement, shop drawings, administration and accounting. In a project requiring design and build, design works, issue of drawings, and revision of plans can constitute a large part of indirect costs.

The following types of cost may be considered in the category of off-site overheads:

1. Executive and clerical salaries.
2. Office occupancy costs (rent, mortgage, services etc.).
3. Design office overheads and testing facilities.
4. Plant workshops, yards and storage areas.
5. General maintenance and depreciation of plant.
6. Advertising, marketing and general administrative costs.
7. Professional fees.
8. Off-site vehicle expenses, office supplies and taxes.

Labour and Equipment. It may be necessary to redeploy resources from other contracts or engage subcontractors at higher rates. Redeployment of labour and sometimes equipment from one part of the site to another may also cause secondary disruption. The test for inclusion of labour and equipment in a delay and disruption claim is whether or not the contractor was present on the site and idle for a longer time due to the delay; or whether the contractor is one who is engaged on a specific work activity and whose time on-site is governed by the rate of progress of that activity.

Subcontractor's Delay Costs. They include nominated, designated and selected subcontract delay costs. Subcontractors may well have an entitlement to delay costs from a contractor. In turn the contractor may consider that such delays were caused by the client and forward these claims to the client. Subcontractors delay claims may include delay costs due to the actions of the contractor and the client. The client must ensure that these claims are separated and must insist that the contractor claim does so. Contractors are often reluctant to provide such a break-up and endeavor to recoup all of a subcontractor delay costs from the client.

It often happens that in the subcontract that there is no specified program and the subcontractor has agreed to carry out the subcontract work in accordance with the requirements of the progress of work under the main contract. In that event it may be extremely difficult to prove that a particular act caused delay, since there would be no base line program to measure delay. To evaluate such claims, it is necessary to have the contractor establish the justification and amount of any claim in the same manner that would be required for the contractor own claims, have a copy of the subcontracts between the contractor and the subcontractors, to check what claims are justified and importantly to have the contractor certify that the subcontractor has been paid the entitlement due or give a direction for the client to pay the subcontractor.

Financing Costs. This is an area where claims have little or no justification. Sometimes, however, a contractor can genuinely incur an extra cost for financing charges. Note that if some delay costs, such as payments to suppliers, may not have been paid out by the main contractor at the date of a payment claim and hence overdraft interest could not be properly included in the claim.

If financing charges are claimed, the contractor should be required to provide proof of the dates upon which amounts were allegedly paid out. Bank statements with payments identified may assist. Proof of payment of interest by the contractor should also be requested. The account may be in overdraft and it may only be necessary to have proof of the overdraft rate. Interest from the date when the money was paid out until the next date thereafter for a payment claim is then calculated.

Claim Preparation Cost. The cost of preparation of a delay claim is not payable as a separate head of damage. The cost is part of the overheads. A contractor is not entitled to additional payment because the contractor chooses to make a separate claim or to use a claim consultant. Just as a contractor is not entitled for cost in order to prepare claims, the client is not entitled to recover from the contractor the additional cost of reviewing claims, even if they are totally without substance.

Loss of Profit and Opportunity. In respect of claims arising from breach of contract, profit is usually not payable. The law permits an award of the actual loss arising from a breach of contract, but not profit on the loss. Loss of profit may be recoverable in rare instances, as part of a claim for loss of income. The loss of income is income from sources other than from the contract breached by the client. The breach by the client must be shown to have prevented the contractor from earning income which the contractor would have earned if the breach had not occurred. Provided that the contractor has mitigated the loss, there may be a permanent loss in income which is recoverable from the client as damages.

It is irrelevant whether the income lost is profit or overheads or categorised in any other way. It only has to be shown to be income lost. In such instances, the real cause of the loss of income is that the contractor is tied up by the delays and that the delays had prevented the contractor from earning income. If, however, the client pays for the contractor resources during the delay period, the lost income is reimbursed through those payments and there is no basis for a claim.

The client would argue that a contractor income will vary from time to time and although the delay on one contract may reduce the income from that contract in one financial year, the income from that contract will be received in the next financial year and there is no permanent loss of income.

Inflation. Delay and disruption causes works to be carried out at later times than originally programmed. Consequently, any increase in labour and materials cost may be recoverable due to the increased costs of these items due to inflation or if special items of materials, such as copper, steel and timber, have increased in prices in the non-culpable delay period due to such increases in the international market.

5.9 Methods for Calculating Claims

There are various methods to quantify loss of productivity in a delay and disruption claims. These methods, each having its own merits and limits, include:

The industry standard approach. The industry standard method of quantifying labour productivity is not always favoured by courts due to its hypothetical nature. Trade groups for various contractors have published productivity tables that show how various job-site conditions can affect labour productivity. Particular conditions are scored in terms of their effect on productivity. A contractor can use, for instance, certain published factors and ascertain how much the contractor productivity was affected by the employer disruption. These measurements are then quantified to show the loss of efficiency and the disruption claim amount. Clients contend that industry standards are factually distinguishable from the conditions actually experienced on their job sites. Nonetheless, industry standards have been used successfully as a way to measure productivity when actual data is missing or unavailable otherwise.

The total cost and the modified total cost methods. The total cost method is based on the premise that the resulting project is a cardinal change from what was originally contracted, that is, the current project is fundamentally different from the project envisioned by the contract or when other methods of computing losses cannot be done. Once a cardinal change, or abandonment, has been established, the contractor is freed from the terms of the contract and is allowed to recover the reasonable value of labour and materials, plus reasonable mark-ups for overhead and profit, less than what was previously paid.

The modified total cost method alters the total cost method by subtracting from the total costs any costs incurred by the contractor due to its own inefficiencies.

The main criticism of the total cost and modified total cost methods is that they can be used by contractors to hide losses not caused by the client, such as those losses due to the contractors' tender errors or defective project management. Clients frequently seek to bar the total cost and modified total cost methods on these grounds. Clients also contend that the contractor cost records are sufficiently detailed so that, during the course of a project, the contractor could have tracked its costs to show an actual causal relationship between the client actions and the contractor loss of production. Clients argue that, to the extent the contractor failed to adjust its accounting system to track job costs, the contractor should be barred from using these methods. The total cost and the modified total cost methods are dealt with much detail in the rest of this book.

The jury verdict method. This method allows the '*trier of fact*'[25] to determine recoverable delay and disruption damages. To apply this method, a party must present:

1. A clear proof of injury.
2. An indication that there is no more reliable method of computing damages.
3. Evidence of sufficient weight of fact to make a fair and reasonable approximation of damages.

Frequently, courts use evidence submitted in support of other quantification methods, such as the methods described above, to derive a jury verdict calculation. Clients contend that the jury verdict method should not be used because the plaintiff contractor failed to meet its burden of demonstrating damages with reasonable certainty. Clients also contend that the jury verdict method, in essence, amounts to nothing more than allowing the jury or the court to make a guess regarding what the contractor damages are, and this process frees the contractor from its normal and customary burden of showing breach, causation and damage.

The measured mile analysis. The measured mile approach, which involves comparing the actual cost with what the cost would have been was it not for the disruption, is one of the main proven methods a contractor applies in order to quantify costs once it has been established that loss of productivity has occurred and caused delay. It involves the evaluation of disruption carried out on the basis of a comparison of productivity prior to the disruptive events taking place, compared with that achieved during the period of disruption. A comparison of outputs is usually made by assessment of sums certified within interim payment certificates. The productivity rate for a period of disruption is quantified in lost worker hours, which are multiplied by an hourly rate to find the loss of labour productivity, or the disruption claim amount.

The measured mile, in its purest application, measures two different periods of productivity for the same type of work performed under the same type of physical conditions on the same project. On this basis any resulting disruption claim will be

---

[25] In a jury trial, the jury is the '*trier of fact*'. In a non-jury trial or a bench trial, a judge, or panel of judges, is the '*trier of fact*'.

in respect of actual loss and expense incurred instead of reference to tender allowances. Clients may disagree with a measured mile calculation by claiming that the baseline productivity measure, that is the measure of the period unaffected by disruption, is faulty. A baseline productivity measure can be inaccurate if the delayed period versus the planned period comparison is not an accurate comparison.

In the case of *Whittal Builders Company Ltd v Chester-Le-Street District Council*,[26] the court had to decide the most appropriate method of evaluating the disruption which had resulted from the problems relating to the handover of the properties. Of the variety of methods presented before the court, the measured mile method of evaluating disruption was the favourite method approved.

In *Whittal Builders,* Mr Recorder Percival had this to say:

> Several different approaches were presented and argued. Most of them highly complicated, but there was one simple one—that was to compare the value to the contractor of the work done per man in the period up to November 1974 with that from November 1974 to completion of the contract. The figures for this comparison, agreed by the experts for both sides were £108 per man week while the breaches continued, £161 per man week after they ceased. It seemed to me that the most practical way of estimating the loss of productivity and the one most in accordance with common sense and having the best chance of producing a real answer was to take the total cost of labour and reduce it in the proportions which those actual production figures bear to one another—i.e. by taking one-third of the total as the value lost by the contractor. (This roughly being the difference between the productivity when work was not disrupted i.e. £161 per man week and £108 per man week when work was disrupted). I asked both (counsel) if they considered that any of the other methods met those same test as well as that method or whether they could think of any other approach which met them better than that method. In each case the answer was 'no'. Indeed, I think that both agreed with me that that was the most realistic and accurate approach of all those discussed. But whether that be so or not, I hold that it is the best approach open to me, and find that the loss of productivity of labour and in respect of spot bonuses which the plaintiff suffered is to be quantified by adding the two together and taking one-third of the total.

In the case of *John Doyle Construction Ltd v Laing Management (Scotland) Ltd.*[27] the court was asked to express a view as to whether a measured mile approach in principle is acceptable. A part of the claim related to disruption and the evaluation was produced by comparing labour productivity actually achieved on-site when work was largely free from disruption with labour productivity achieved when work was disrupted. It was decided that this method of evaluating a claim was not a total cost claim and in principle was acceptable.

---

[26] (1985) 11 CLR 40. The facts of the case arise from a contract let by Chester-le-Street (the defendant) to Whittall Building Company Ltd (the plaintiff) for the refurbishment of 90 dwellings. Difficulties were experienced by the defendant in granting possession of the properties. The court found that during the period when these problems existed the contractor was grossly hindered in the progress of his work and as a result ordinary and economic planning and arrangement of the work was rendered impossible. However a stage was reached in November 1974 when dwellings were handed over in an orderly fashion and no further disruption occurred.

[27] (2004) BLR 295.

5.10 Quantifying Delay Claims

Delay claims are claims for additional time-related costs associated with delays caused. Therefore, a delay claim is a claim for delay costs where a contractor must show that the cause of the delay is one that entitles the contractor to payment for the extra costs incurred. To establish an entitlement to delay costs, a contractor must demonstrate that a delay to completion of the contract has occurred and must show that the cause of the delay is one that provides the contractor with an entitlement to extra payment, either under a term of the contract or for breach of contract by the client. The claim must be supported by evidence of the facts on which it is based. To warrant the payment of delay costs, a delay must affect the critical path and delay completion of the whole of the works or a milestone. If claims for delay costs are not handled appropriately there are risks that excessive costs will be incurred or contract disputes will occur.

When a contract specifies rates for delay costs that are applicable to the events causing the delay, calculating delay costs is straightforward once the appropriate extension of time has been determined. The contractor is entitled to be reimbursed at the specified delay costs rate for the period by which the contract time was extended, subject to any exclusions stated in the contract. When there is no specified rate for delay costs, calculating a contractor entitlements can be complex and time consuming.

To quantify a delay claim, a contractor must demonstrate that the delay caused damages, in the case of a breach of contract, or extra cost, in the case of a specific contract provision. The onus is on the contractor to prove that the costs claimed have been incurred and that every effort has been taken by the contractor to minimise these costs. If a contract does not include a prescribed delay cost rate, then it is necessary to assess what delay costs are legitimate and to evaluate those costs.

Various methods of calculating a delay have been adopted which vary according to the wording of each contract, the circumstances of the delay and the experts involved in assessing the claim. In essence, what is being sought is the delay which is caused by the relevant events occurring at time which it in fact occurs, with the project being in the state which it is in at that time and the contractor responding to it as he does, with an allowance, then, being made for any extent to which the contractor has, through breaches of contract, contributed to the resulting delay. Such contribution might have affected the state of the project at the time the delaying event occurred or might have affected it afterwards if, for example, the contractor failed to comply with an obligation to use his best endeavours to overcome a delay.

For lawyers, the underlying question is whether or not the tribunal will be convinced, on a balance of probabilities, that a relevant event, rather than other events, which caused delay of certain duration. Models may be sufficient to demonstrate it but, equally, they may not. Sometimes a theoretical model will be acceptable if the conditions for its reliable use are satisfied or, even if there is doubt over this, it is accepted for reasons of economy or practical necessity in the context of the dispute. Mostly, proof of delay is a matter of fact but sometimes facts alone cannot answer the question and the law is required to take a position.

5.11 Methods for Delay Assessment

The various approaches used by programming specialists to demonstrate or assess the delaying effect of particular relevant events include:

Bar Chart Methodology. Contractors seeking to prove entitlement to time or money often use a bar chart comparing the planned and as-built programme. The contractor will then claim relief for the difference between the end dates shown by the two programmes. The bar chart methodology demonstrates excusable delays and provides that certain conditions are satisfied namely that:

1. The planned programme does not contain any float.
2. All the events giving rise to the delay can be clearly identified and are excusable events.
3. There is no need to take account of concurrent or parallel delays or accelerated or inefficient working.

Retrospective Critical Path Method. This analysis is relatively new and has revolutionised the way construction projects are programmed and managed and on the way the effect of delay can be predicted and calculated. Its main advantage is the opportunity which it provides to link cause and effect at a level of considerable detail. There are several variations on the retrospective critical path method analysis including:

1. *As planned impacted* which adds client caused delays into the as-planned programme.
2. *As built but* which subtracts employer caused delays into the as-planned programme.

The two predominant areas of retrospective delay analysis are static and dynamic. Static critical path analysis is largely inferior to a dynamic critical path analysis and is usually adopted when the cause is clearly identifiable; there are no complex issues such as acceleration or unproductive work and there is no change in the logic. However, because construction contracts are dynamic that is, the critical path will change, dynamic critical path methodology is the preferred route to retrospective delay analysis. Dynamic critical path methods are classed as time-impacted analysis and are based upon the analysis of delaying events at the time they occur. However, what method is adopted will be decided upon by factors such as proportionality and what materials are available in order to construct an as-built programme.

Window Analysis. The construction process is seen as multiple windows during the period of performance. For each window the programme is updated to take account of delays which are at the contractor risk, any necessary logic or duration revisions by of mitigation and all excusable and compensable events during the period since the last update. Thus, each programme update will incorporate all changes which affected the planned progress of the project during that period. This method is most effective when used contemporaneously and is updated regularly throughout the course of the project. Since each window is only a segment of the contract period, the result of each window analysis must be summarised and carried forward to the next window. It is, thus, only when the last window, closing at actual completion, has been analysed and summarised that the accumulation of the various changes can be added together to demonstrate the effect on completion of the various contingencies. In simple terms, what is being constructed is an as-built programme which shows the impacts of all delaying events as they occur.

The window analysis is the best proven technique for determining the amount of extension of time that the contractor should have been granted at the time that an excusable risk occurred. It is also quicker and easier to apply as in each window where there are relatively few activities to be analysed, as compared to the overall programme. However accurate progress information at the time of the windows must be available, otherwise the analysis cannot be properly or accurately completed. The less accurate the programme and progress information available is, the more likely that results will be obtained that is clearly inaccurate, and that will require to be amended by manipulating any obvious errors in the original as planned programme.

Snapshot analysis. The snapshot analysis it is the occurrence of the event itself and the cessation of its operation which dictates what is analysed and at what point in the progress of the project it is analysed. This is a simple contemporaneous approach to delay analysis which allows assessment to be made of three important aspects:

1. The actual state of progress at the time the delaying event occurred.
2. The changing nature of the critical path as a result of the events.
3. The effects of action taken, or which should have reasonably been taken, to minimise delays or avoid subsequent delays.

5.12 Delay Claims Methodology

On projects where the level of construction activity varies significantly between various stages of construction, the appropriate costs will be those which relate to the periods in which delay has occurred. To adopt average daily or weekly site overhead rates over the whole contract duration is not necessarily appropriate. The weekly cost of site overheads is related to the total work activity being undertaken on a construction site at any time and not only to work on the critical path. Although a delay to critical work may occur at a period of maximum site activity, the effect may be to prolong only relatively few activities for additional time.

The appropriate rate for calculation of site overheads is that related to the sequence of activities which were delayed, together with any consequential effects attributable to the delay.

Some non-critical activities will be delayed by a delay to critical activities on which they were dependent. They may still be non-critical, but they will be undertaken at a later time. This shift in time may not result in additional site overheads overall. It may merely cause the overhead costs to be incurred at a later time. Site overheads are broadly related to the direct cost of the work to be undertaken. For projects with a high-labour content, the cost of supervision will be greater than for projects with a high plant or material content. Less supervision is required for work where subcontractors are used compared with work requiring unskilled labour or using the contractor employees. The cost of engineering staff on civil engineering or complex multi-disciplinary projects may be higher than on building contracts.

One method of assessing on-site delay costs is to evaluate the actual costs incurred. If cost information is provided by a contractor to justify a claim, then this must be audited to eliminate all costs that should be included in the direct costs of construction activities. Generally, there will be very few material costs to site overheads. Costs that are not time related would be deducted. For example, mobilisation and removal costs will be incurred irrespective of the contract duration. The exception could be where equipment or staff may be demobilised temporarily at the beginning of a long delay and remobilised at the end of the delay period in order to save the cost of retaining these resources on the site for the duration of the delay period.

The actual cost of staff and labour will generally be provided by wages or salary sheets and the cost of all external plant and services will be substantiated by invoices. These invoices should be inspected to ensure that no operating charges such as repairs or replacement parts, fuel or other incidentals are included. The delay cost is the equipment rental charge only. In some cases, there will be reduced hire rates or even no charge for standby or non-operational periods. Where contractor owned plant is involved, invoices are not likely to be available, so an analysis of the costs claimed will have to be made separately.

The importance of site records to assist in recognising excessive costs cannot be overstated. Where the client has maintained good site records, the checking of the actual times claimed against a contractor records is the ideal way of establishing costs and the best way of accurately assessing the costs. A contractor is required to mitigate costs in the event of delay. Excessive expenditure due to poor management or inefficiency, if proven, is not recoverable. Specific instances of obvious waste or mismanagement would need to be identified in the costs or noted from site observations.

Examples of actions that could be taken by contractors to minimise delays impact on the overall completion of a project and to mitigate losses, include:

- Terminating hire of plant not being used or hire out plant.
- Laying off workers who are not productive when this is possible.

- Adding on additional workers or subcontractors, with a net benefit of a reduced delay impact.
- Working overtime.
- Re-organising work programs and order of works.

5.13 Quantifying Disruption Claims

A disruption claim is a claim for alleged disruption or loss of productivity resulting from the acts or omissions of the client or the client agents. Disruption claims are sometimes included with delay claims but are fundamentally different. Disruption may not result in delay. In a disruption claim, contractors claim that they could not achieve their planned output because of the client actions and hence the damages or extra costs are payable. Disruption in its simplest sense is hindrance to actual progress thus reducing the output of construction resources, those being, primarily labour and plant. Contractors claim that they could not achieve their planned output because of the client actions and hence the damages or extra costs are payable.

Disruption, also called loss of productivity, results in a delay to the work being carried out and not necessarily to the completion of the works. The work produced is not changed, it simply takes longer time to complete. A contractor may claim disruption costs independent of an extension of time claim, especially where the contract makes provision for this. To warrant the payment of disruption costs, a contractor must identify the particular work activities that were affected by the disruption or loss of productivity and must demonstrate that the disruption caused the contractor to incur additional costs. The fact that there are many variations, no matter how many, does not give rise to an entitlement for a claim for loss of productivity. For a disruption claim to be valid, a contractor must also demonstrate disruption to actual progress, not planned progress as is often claimed. A work as executed program can be the starting point for any demonstration of reduced productivity.

An example of disruption is where the client ordered a contractor to cease work on a particular activity for frequent short periods (for example to provide for some necessary operating function of a plant or building) and the need for such stoppages was not specified in the tender documents. Another example would be where the client ordered urgent variations and groups of workers had to continually move from one activity to another at short notice, being unable to develop optimum productivity on a particular work activity. Hence, a disruption claim must identify specific events that are breaches of contract by the client or events for which the contract specifically provides for extra costs.

A contractor must quantify the disruption costs once it has established that loss of productivity has occurred and caused a delay. This involves comparing the actual cost with what the cost would have been was it not for the disruption. The contractor must demonstrate that the latter cost is reasonable, although it is hypothetical to some extent. In making such claims, a contractor must also establish that everything reasonable has been done to minimise the cost of the disruption, for example hired machines were not left idle on-site when the hire could have been terminated. Cost details of the affected resource may then be compiled from the site accounting records. On this basis any resulting disruption claim will be in respect to actual loss and expense incurred instead of reference to tender allowances.

5.14 Disruption Costs Methodology

The calculation of disruption costs flows from the extensions of time granted on account of delays caused by the disruption. Where a contract does not prescribe a method for evaluating disruption costs, a contractor might make an ambit claim for the difference between the amount allegedly allowed in the tender and the actual cost of the work performed. Such an approach falsely assumes that the amount allowed for the work at the time of tender was totally correct and there were no inefficiencies in the contractor's management of the construction operations.

A contractor must quantify the disruption costs once it has established that loss of productivity has occurred and caused a delay. This involves comparing the actual cost with what the cost would have been if not for the disruption. The contractor must demonstrate that the latter hypothetical cost is reasonable. In making such claims, a contractor must also establish that everything reasonable has been done to minimise the cost of the disruption, for example hired machines were not left idle on the site when the hire could have been terminated.

Disruption losses are commonly established by the measured mile technique. This compares productivity achieved on an unimpacted part of the contract with that achieved on the impacted part. The following represent the typical heads of claim for recovery in disruption claims:

- Indirect costs and head office overheads–easier to establish this as a cost incurred if charged to the job. If not charged to the job, costs may be allocated on a proportional basis.
- Labour and equipment–uneconomical use of labour and plant, idling time and whether the equipment is owned or on lease.
- Loss of profit–lost profit on other contracts is generally not recoverable under the standard forms. However, where there are no limiting contract provisions, a claim for loss of profit under the general law may be possible.

5.15 Mitigation in Construction: A General View

The contractor has a general duty to mitigate the effect on its works of the client risk events. This duty to mitigate does not extend to requiring the contractor to add extra resources, or to work outside its planned working hours, in order to reduce the effect of an employer risk event, unless the employer agrees to compensate the contractor for the costs of such mitigation. However, it can be argued that the obligation to progress the works diligently may, however, requires the contractor to take some positive action, and a failure to do so may sound in damages measured by the liquidated damages for additional period of overrun which could have been avoided but for the breach.

Generally, a contractor has a duty to mitigate delay impact if practical. In determining what mitigation is practical the courts consider:

(1) Whether the delay is of a reasonably known length to allow planning.
(2) Whether there are other projects, existing or new, that can use these resources effectively during the delay.
(3) What the costs of reassigning the resources will be.
(4) Whether the delayed project can be partially or totally demobilised or utilising space sacrificed.
(5) How the subcontractors and suppliers will be affected and how the delay impact on them can most effectively be managed.
(6) Whether the remaining work can be re-sequenced to allow real progress to be made during the delay, and the cost of that re-sequencing.

The remedy of an extension of time is a contractual remedy for acts of prevention and breach of contract by the client and for events at the risk of the client. It may, therefore, be thought that if the remedy of extension of time is based on causation, then the principles referred to as the duty to mitigate should apply. It is suggested that there are two situations to consider, first when the contractor responds positively and the second when the contractor takes no positive action.

In the first situation the contractor may react to the qualifying delay by making changes to his methods of working or sequence of working or even to accelerate the work. The issue, then, is whether he is entitled to recover the loss incurred by this reaction and that depends on whether or not he has a right to react as he did. It is suggested that subject to the express terms of the contract, the contractor has no right to accelerate and is not entitled to recover additional costs incurred in acceleration measures to mitigate the effect of qualifying delays without an instruction from the client.

Since many contracts contain provisions for the grant of extensions of time and express terms for agreement of acceleration measures, the unilateral action by the contractor in giving priority to the fixed date for completion over the cost of working efficiently cannot bind the client in those contracts. It is suggested that this interpretation can be expressed in terms of the reasonableness in mitigation. It is not reasonable when there are sufficient contractual remedies for the contractor to decide to accelerate the works. This interpretation must be examined in the context of express obligations to progress the works.

In the second situation the contractor may not react to the qualifying delay and the issue then is what minimum measures he is required to take in order to mitigate the effects of the qualifying delay and if he fails to take those measures whether this affects the extent of his entitlement to extension of time. It is suggested that although the rules of mitigation do not generally apply to construction contracts with extension of time provisions and provision for recovery of time related losses, the contractor will have some obligation to progress the works which will involve an aspect of management of resources and planning of activities in the circumstances of actual events. Although as a matter of interpretation of the terms of the contract, it is suggested that such an obligation will usually be intended by the parties to apply equally to events causing qualifying delays.

5.16 Minimising Disputes: Practical Steps

Disputes arise generally when there is an irreconcilable difference of opinion between the parties over their obligations. The number of disputes could be substantially reduced by the introduction of a transparent and unified approach to the understanding of elements of disputes such as the concept of time, liquidated damages, time at large, completion, concurrency and programmed works.

To reduce the number of disputes relating to delay, the contractor should prepare a programme showing the manner and sequence in which the contractor plans to carry out the works. The programme should be updated to record actual progress and any extensions of time granted.

As to the ownership of float, the parties should expressly address the issue in their contract. The ambiguous interpretation of total float ownership can be clarified by improving contract language with regard to specifications in the area of total float management. The proposed concept for managing total float should involve pre-allocating a set amount of total float on the same non-critical path of activities to the two contractual parties—owner and contractor. Given the difficult nature of this issue, this seems as an inherently sensible recommendation.

While this apparent conflict found many leading cases on float for the proposition that there is no implied term in building contracts that the client should perform the contract so as to enable the contractor to complete the works in accordance with a programme showing a completion date earlier than the contractual completion date; as a matter of policy, contractors ought not to be discouraged from planning to achieve early completion because of the price advantage that being able to complete early is likely to have for the client. Accordingly, if a client delay prevents the contractor from completing at an earlier date than the contractual completion date, the contractor should, in principle, be entitled to be paid the direct costs of the client delay. This is made subject to the significant proviso that the client was made aware of the intention of the contractor to complete early prior to entering into the contract.

The parties should adopt, wherever possible, the practice of pre-agreeing the total likely effect of variations, so that there is a fixed price of the variation to include the direct costs of labour, plant and materials and time-related costs. Where it is not practical to agree in advance the amounts for delay and disruption to be included in variations and sums for changed circumstances, then it is recommended that the parties to the contract do their best to agree the total amount payable as the consequence of the variations and/or changes separately as soon as possible after the variations are completed.

It is to be noted that the courts place particularly heavy burden upon contractors in terms of the maintenance and presentation of documentation in support of any claim for delay. The contractor has to maintain accurate and complete records, and should be able to establish the causal link between the client risk event and the resultant loss and/or expense suffered.

Where the contract provides for specific procedures to be followed as a condition precedent to the valid exercise of the power to deduct liquidated damages, for example, a written application from the contractor or a certificate issued by the contract administrator, such a procedure must be complied with. Whether the giving of a notice is a condition precedent is a question of construction of the particular contract.

Liquidated damages are an increasingly unattractive means of providing incentives for increased performance because they run the risk scrutiny by the courts determining that they are a penalty, particularly where much of the loss which the client envisages suffering is non-pecuniary and the liquidated damages are included more as a management tool than anything else. They may be regarded as inconsistent with trends which are more towards the use of positive incentives of good performance rather than punishment for bad performance.

This raises interest in the use of various forms of bonus agreement; particularly those pain/gain deals which envisage that the contractor will earn more for earlier planned performance just as much as he may earn less for later planned performance.

References

Case Law

Bovis Construction (Scotland) Ltd. v Whatlings Construction Ltd. (1995) 49 Con LR 12
Bramall and Ogden v Sheffield City Council (1985) 1 Con LR 30
British Steel Corporation v Cleveland Bridge & Engineering Co. (1981) 24 BLR 100
Commissioner of Public Works v Hills (1906) AC 368
Dunlop Pneumatic Tyre Co. v New Garage and Co. Ltd. (1915) AC 79Motor
Hadley v Baxendale (1854) 9 Exch 341
John Doyle Construction Ltd. v Laing Management (Scotland) (2004) BLR 295
McRae v Commonwealth Disposals Commission (1951) 84 C.L.R
Moschi v Lep Air Services Ltd. (1972) 2 All ER 393
Occidental Worldwide v Skibs A/S Avanti (the Siboen and the Sibotre) (1976) 1 Lloyd's Rep. 293
Pavey and Matthews Pty Ltd. v Paul (1987) 69 ALR 577
Peak Construction (Liverpool) Ltd. v McKinney Foundations Ltd. (1970) 1 BLR 111
Philips v Ward (1956) 1 WLR471
Philips Hong Kong v A-G of Hong Kong (1993) 61 BLR 41

RP Wallace Inc. v The United States (2004) 63 Fed Cl 402
Regional Construction Sdn. Bhd. v Chung Syn Kheng Electrical Co. Bhd. (1987) 2 MLJ 763
Robinson, Administrator of Robinson v The United States (1922) 261 US 486
Robophone Facilities Ltd. v Blank (1966) 1 W.L.R. 14
Sabemo Pty Ltd. v North Sydney Municipal Mutual Council (1977) 2 N.S.W.L.R. 880
Sun Shipbuilding and Drydock Company ASBCA 11300, 68-1 B.C.A. (1968)
Teacher v Caddell (1889) 1F (HL) 39
United States v United Engineering Co. 234 U.S. 236 (1914)
Whiten v Pilot Insurance Co. (2002) S.C.R. 595
Whittal Builders Company Ltd. v Chester-Le-Street District Council (1985) 11 CLR 40
William Lacey (Hounslow) v Davis (1957) 1 W.L.R. 932

Books

(2002) Delay and disruption protocol. Society of construction law, UK
Beale H, Bishop W, Furmston M (2008) Contract: cases and materials. Oxford University Press, Oxford
Bockrath J, Plotnick F (2010) Contracts and the legal environment for engineers and architects. McGraw-Hill, NY, USA
Bramble B, Callahan M (2010) Construction delay claims. Aspen Publishers, Maryland
Burrows (2005) Remedies for torts and breach of contract. Oxford University Press, Oxford
Callahan M (2010) Construction change order claims. Aspen Publishers, Maryland
Carnell N (2005) Causation and delay in construction disputes. Blackwell Publishing, Oxford
Cushman R, Carter J, Gorman P, Coppi D (2007) Proving and pricing construction claims. Aspen Publishers, Maryland
Davison P, Mullen J (2008) Evaluating contract claims. Wiley-Blackwell Publishing, Oxford
Eggleston B (2008) Liquidated damages and extensions of time: in construction contracts. Wiley-Blackwell Publishing, Oxford
MacGregor H (2010) MacGregor on damages. Sweet and Maxwell, London, UK
Pickavance K (2010) Delay and disruption in construction contracts. Sweet and Maxwell, London, UK
Reese C (2010) Hudson's building & engineering contracts. Sweet & Maxwell, UK
Schwartzkopf W, John J, McNamara J (2000) Calculating construction damages. Aspen Publishers, Maryland
Wickwire J, Driscoll T, Hurlbut S, Groff M (2010) Construction scheduling: preparation, liability, and claims. Aspen Publishers, Maryland

Chapter 6
Global Claims: An Overview

6.1 Global Claims: Definition

The context in which the term global claim is most frequently used is to describe a contractor's claim for delay or for a loss resulting from a number of different causes for which the client is responsible and the claimant does not seek to attribute any specific loss to a specific breach of contract, but is content to allege a composite loss as a result of all the breaches alleged, or presumably as a result of such breaches as are ultimately proved. The reason for making a global claim is that due to the complex interaction between two or more delaying or disruptive causes or events, it is impracticable or even impossible to accurately apportion a particular sum to a particular effect and to a particular cause.

A global claim has been defined, as the name suggests, as a global or composite sum put forward or claimed as damages due to two or more separate heads of claim or events, where it is alleged that it is impracticable or impossible to provide a distinct sum claimed for each of the cause and effect. A global claim is also known as a composite, rolled-up and loss and expense claim. In the United States it is better known as a total cost claim. As global claims do not involve linking individual breaches to individual losses, they are easier to prepare and can be presented in a more commercial manner than traditional legal claims. When used at an early stage in a dispute, they can result in a quick settlement that avoids the need for expensive case preparation.

A global claim is a claim where the composite loss is often prepared as a total cost claim where the quantification of loss is achieved by subtracting the tender cost of the works from the final cost. It is a claim where the claimant has the responsibility to adduce evidence to prove the essential elements of the global claim such as a breach of contract, causation and the burden of loss and offers a collection of events and breaches of the total sum of loss incurred and asserts that the former caused the latter.

Global claims are a modification of the basic principles of contractual claims, so far as the courts will allow such modification. A global claim is permissible

A. D. Haidar, *Global Claims in Construction*,
DOI: 10.1007/978-0-85729-730-3_6, © Springer-Verlag London Limited 2011

where it is impractical to disentangle that part of the loss attributable to each head of claim, and the situation has not been brought about by delay or other conduct on the part of the plaintiff. In such circumstances the court infers that the defendant breaches caused the extra cost or cost overrun and the *causal nexus* was inferred rather than demonstrated.

Hudson[1] defines global claims as:

> A global or composite sum, however computed, is put forward as the measure of damages or of contractual compensation where there are two or more separate matters of claim or complaint, and where it is said to be impractical or impossible to provide a breakdown or sub-division of the sum claimed between those matters.

There are mainly two types of global claims, namely:

- Loss and Expense. The claim is usually based on an allegation that there were numerous variations in the contract and the costs overran. The claimant, then, alleges the cost overrun is recoverable as a result of the variations. There is, however, no analysis that a particular variation leads to a particular item of cost.
- Delay and Disruption. The claim is usually based upon an allegation that there were numerous variations events interfering with the works and the works were delayed, entitling the contractor to an extension of time and monies. Again there is no link between the alleged events and delay. There are a number of ways in which the global sum may be quantified, but it is usually done on the basis of the total additional cost said to be the result of the matters complained of.

A strong and unspoken point of a global claim is that it pushes the parties to settle. A global claim is sometimes used as this bargaining tool but one has to be careful not to use a global claim as a blackmail or ransom tool. Therefore, when submitting this type of claims, one should be as fair and reasonable as he can and be clear that, although willing to settle, the basis of the claim are just and the claimant is crystal clear that the components of the global claim exist.

6.2 Basic Principles

In a construction contract context, it is well-settled law that the claimant has the evidential burden of establishing that:

(1) A breach of contract (duty or other claims event) has in fact occurred and that the defendant is factually and legally responsible for it.
(2) The breach (claims event) caused the loss alleged to have been suffered.
(3) A loss has been suffered and the quantum of that loss cannot be itemised by determining the amount of the loss per each causative event.

---

[1] Hudson's Building & Engineering Contracts.

The correct manner of presenting a claim is to link the cause with the effect. However, this is not always easy, especially when the claim is a disruption claim rather than a prolongation claim. To counter such difficulties, contractors have, in recent times, attempted to shortcut the need to link cause and effect by the use of the global claim. A global claim is, therefore, a disguised delay and disruption claim where the causes are rolled up or joined to form a global effect with the resulted overrun cost as the damage claimed and pursued by the injured party.

The main principles of a global claim are:

- Breach/claims event. The claimant has the burden of proving that the breach of contract, breach of duty or other claims event actually occurred and that the defendant is factually and legally responsible for all of the reasons the claim is submitted forth for.
- Causation. The claimant has the burden of proving that the breaches caused all of the losses alleged to have been suffered due to the breach of contract of the other party.
- Quantum. The claimant has the burden of proving loss suffered and amount of that loss. The amount of loss is usually the total cost that the claimant has incurred over and above he priced the works for. It is the difference between the actual cost of the contractor and the cost he tendered for in his bid.

Unsurprisingly perhaps, the fundamental objection to global claims is that they unashamedly offend the generally accepted legal position. In particular, no case on causation is advanced because the sums claimed are not sought to be linked to individual breaches, claim events or causes of action. In such claims, the claimant simply takes the difference between the total actual costs and the total estimated costs of carrying out the works as being the increase in costs representing the damage suffered due to the delaying or disruptive events. There is no *nexus* or connection shown between the individual events to their consequences whether in terms of time or money claimed. The contractor simply may plead a lengthy list of delay inducing variations, allegations of denial of access, late receipt of critical information, interference by other contractors or statutory undertakers.

A global claim may be viewed as an exception to, or at least a modification of, the generally accepted position, since the claimant openly declares his intent not to adduce evidence to prove the basic elements but rather puts forward a collection of breaches and events and a total amount of loss incurred and asserts that the former caused the latter.

Furthermore, global claims tend at worst to ignore and at best to camouflage all other alternative reasons as to why the total costs exceeded the original contract sum; for example, unrealistically low tender price, inclement weather, labour shortages, lack of proper documentation and planning and management inefficiencies.

Global claims, where a loss is attributed to a list of events with no specific link to each part of the claim, can be particularly useful when dealing with the complicated processes of construction. When a party to a contract makes a claim to recover a loss, it is normally required to prove a connection between individual events and each item of loss. But the process of construction is a complicated interaction of activities and the overall loss might well be caused by the combination of a number of different events. If this is the case, and it is impossible to trace the connection between each individual event and each individual loss, then a global claim is often made.

On the other hand, proof that an event which played a material part in causing global loss but was not the responsibility of the other party would undermine the logic of the global claim. In addition, if the other party proved that other factors, for which it held no liability, had made a material contribution to the cause of the global loss, then the global claim would again be undermined.

Courts have found difficulties in handling these cases and their opinion is diversified depending on the facts and the causation principles applied to each case individually. J. Smith in *Nauru Phosphate Royalties Trust v Matthew Hall Mechanical & Electrical Engineering Pty Ltd*[2] said:

> Global claims are difficult for the parties and the court to handle. To compel a plaintiff to give particulars of nexus or justify its inability to do so may reveal the bogus claim. If particulars are produced they may clarify issues and reduce the area of argument even if the plaintiff can only provide alternative hypothesis. I can see no reason why, for example, a judge controlling a building case list or arbitrator could not require the plaintiff to particularise the nexus or to justify its assertion that it is not possible to do so. Such directions would be justifiable upon the grounds that they would assist in the management of the litigation. The issue raised here for decision is whether there is an abuse of process arising from the globally pleaded claim. I consider that, in all the circumstances, there is not.

The global cost method was adopted in the case of *Inserco Ltd v Honeywell Control Systems*[3] where the judge comments on this matter were made:

> Inserco pleaded case provided sufficient agenda for the trial and the issues for the trial and the issues are about quantification. Both Crosby v Portland District Council[4] and London Borough of Merton v Stanley Hugh Leach[5] concerned the application of contractual clauses. However I see no reason in principle why I should not follow the same approach in the assessment of the amount to which Inserco may be entitled. There is here as in Crosby an extremely complex interaction between the consequences of the various breaches, variations and additional works and in my judgment it is impossible to make an accurate apportionment of the total extra cost between the several causative events. I do not think that even an artificial apportionment could be made—it would certainly be

---

[2]   (1994) 2 V.R. 386.

[3]   (1998) EWCA Civ 222. Honeywell were engaged by Olympia & York to supply electrical systems for the Canary Wharf development. They in turn had engaged Inserco to wire up those systems. The subcontract provided for a remuneration on a re-measurement basis. Work commenced in April 1990 but by February 1991 Honeywell were under considerable pressure to complete the project by 1 April 1991. They asked Inserco to provide more labour to enable this to be achieved. Both parties negotiated and agreed that Inserco would be paid on a weekly basis by reference to the men on-site and thus remuneration would no longer be on a re-measurement basis but on a cost plus basis. Subsequently there were disruptions and delays and significant extra works which led to a complex case based on a global claim.

[4]   J. Crosby & Sons Ltd v Portland Urban District Council (1967) 5 BLR 121.

[5]   (1985) 32BLR 51.

extremely contrived—even in relation to the few occasions where figures could be put on time etc. It is not possible to disentangle the various elements of Inserco claims from each other. In my view the cases show that it is legitimate to make a global award of a sum of money in the circumstances of this somewhat unusual case which will encompass the total costs recoverable under the February agreement, the effect of the various breaches which would be recoverable as damages or which entitle Inserco to have their total cost assessed to take account of such circumstances and the reasonable value of the additional works similarly so assessed.

Advancing a claim for loss and expense in global form is a risky enterprise. If the claimant proves that a particular event caused the global loss but fails to prove that the defendant was liable for such event then the global claim will be undermined. However, if the claimant proves that the defendant was liable for a particular event but fails to prove that such event contributed to the global loss then the global claim will not be undermined provided that the claimant proves that the remaining events, for which the defendant was liable, caused the global loss. To the contrary, if the defendant proves that, in addition to the factors for which he is liable, another factor for which he is not liable has contributed to the global loss, the global claim will be undermined.

Although in the circumstances as outlined and when a global claim may be undermined it does not follow that all claims will fail. The fact that the claimant has advanced a global claim because of the difficulty of relating each causative event to an individual sum of loss or expense does not mean that after evidence has been led it will remain impossible to attribute individual sums of loss or expense to individual causative events. In other words, although the global claim may fail, there may be in the evidence a sufficient basis to find causal connections between individual losses and individual events, or to make a rational apportionment of part of the global loss to the causative events for which defendant has been held responsible.

6.3   Limitations

A global claim is often made in a situation where a loss is attributed to a combination of events, without any specific link being made between each part of the loss and each event. All the events that contribute to causing global loss must be the liability of the other party. If an event that played a part in causing global loss was not the responsibility of the other party, this would undermine a global claim.

One of the major objections to global claims is that as pleaded, it contradicts the fundamental principles of pleadings, be it in court or arbitration as it contradicts the principle that the other party must know the case it faces in full particulars so that it is not prevented from raising differing alternatives or particular defences rather than a mere global defence. Such technique of pleading is aptly called the 'forest technique'. The technique raises wide and general terms encompassing all possible eventualities that could arise under the contract without exceptions.

Such a pleading does not inform the other party of the exact nature of the claim made against them being material facts or particulars relied on for such a claim, which the other party is entitled to know. The other party or indeed the court or arbitrator is left guessing as to the details of the claim and may be caught by surprise later. This global pleading also unfairly allows the party to change its course during hearing.

A global claim, in effect, will merely state the list of delaying and disruptive events, for which the respondent is said to be responsible and the global effect of the list of events which may be represented by a global period with the ensuing increased costs as represented by the global sum claimed. Not usually, even the *nexus* between the events claimed and the periods of delay caused by the events are not pleaded but instead a rolled-up period of delay pleaded. Therefore, global claims would unjustifiably place a lax contractor who does not keep proper record in a better position of being allowed to make a global claim as opposed to those complying with the contractual requirements who will not be allowed to make a global claim once shown to have some degree of proper record keeping.

A global claim suffers from a fundamental flaw in which it assumes the claimant as having been perfect and not culpable in any way whatsoever or howsoever for the rolled-up causes to the rolled-up effects and for the increased total cost global claim. In the event if there is any evidence of the probability that actual cost overrun had happened due to the claimant's own risk such as under pricing, poor site organisation, poor costs controls, inexperience or even external factors such as labour strike, labour shortages and inclement weather, then the court is left without any method of gauging the true liability and quantum for the purpose of assessing the true entitlement of the claimant. In fact, it is this fundamental flaw that respondents will be looking to exploit to undermine or even fatally destroy such a global claim.

This simple danger of relying on global claims is that a claimant proceeding on a global claim basis runs an enormous risk of the entire claim being dismissed in the event that liability for any of the causes said to be materially contributing to the global claim is decided against the claimant. This is on the basis that there is no evidence of the make-up of the damages for the other events of claim that may be allowed and as such may entitle the claimant to nominal damages.

Such a form of pleading at face value seems to unreasonably and unfairly switch the burden on to the other party to displace each and every cause and effect without truly being able to judge the entitlement on a cause to effect basis. For instance, allegations of a late issuance of a particular construction drawing may *prima facie* be shown to be late by two weeks after its due date based on the work programme. However, this delay may or may not have affected the progress of the works to be carried which is said to have been delayed by two weeks. Now, if during this period there was another cause of delay raised by the claimant which may not be truly the respondent risk and again the effect of this delay is not globalised to the two weeks, the respondent might analyse the cause of effects occurring to the claimant and raise this delay event as a defence.

In *Hudson Building & Engineering Contracts* the objection to global claims is articulated as follows:

> It is submitted that, in the English and Commonwealth jurisdictions, claims on total costs basis, a fortiori, in respect of a number of disparate claims, will prima facie be embarrassing and an abuse of the process of the court, justifying their being struck out and the action dismissed at an interlocutory stag.

The much publicised *Delay and Disruption Protocol*[6] (2002) takes an extremely firm line in relation to global claims, firmer perhaps than the courts have ever taken. The Protocol states, in effect, that if accurate and complete records are maintained, the contractor should be able to establish the causal link between a client risk event and the resultant loss and/or expense suffered without the need to make a global claim and that the failure to maintain such records does not justify the contractor in making a global claim. If there exists a contractual obligation upon the contractor to maintain particular types of records, perhaps in a defined format with a certain degree of detail, then if a failure to do so results in the pursuit of a global claim, it is not difficult to see that this would be another factor for a tribunal to consider when determining the merits of a global claim.

However, in the post era of *John Doyle v Laing*,[7] it seems that an assertion by the client that a contractor has failed, in breach of contract, to maintain certain records would form the basis of a successful strike-out application. The quality of records maintained is plainly a question of degree and detailed consideration would be required of the records in existence and the reasons for the non-availability of other records. A client might be well advised to insist that, by way of express contractual provisions, a contractor maintains the records recommended. This may limit the necessity and opportunity for making a global claim but it is unlikely to eliminate it entirely.

In *J.Crosby & Sons Ltd v Portland Urban District Council* (1967) (*supra*), J. Donaldson made it clear that global claims were limited to only instances where it is justified because it is impracticable or impossible to make an accurate apportionment of the claim to a particular event but where the individual items of the claim can be dealt with in isolation the arbitrator must make an individual award. He also emphasised that the global claims approach should be used as a last resort method only.

J. Vinelott in *London Borough of Merton v Stanley Hugh Leach* (1985) (*supra*) went on to suggest further limitations to a globally pleaded claim by requiring that the difficulty in apportioning the claim to particular events must not have been created by the claimant's unreasonable delay in making the claim and that the court and tribunals cannot consider themselves obliged to go through volumes of evidence produced by the claimant so as to assess the dominance of events and the effects thereto. In an adversarial system, the onus of proving the case must remain

---

[6]   Society of Construction Law (2002) UK.

[7]   John Doyle Construction Ltd v Laing Management (Scotland) (2004) BLR 295.

with the claimant. The evidential burden of proof should not be simply passed onto the respondent by the mere cry difficulty on the part of the claimant.

He has put, *obiter dicta*, proviso for instance how to deal with global claims such that for a claim to be acceptable only in that the evidence of particular cause to effect to damage had been established without a doubt. However, to this acceptance another condition must be added, in that if the plea was global but the evidence led gave rise to distinctions, the respondent must be entitled to time to consider the development of the evidence for which he may not have previously been aware or given fair warning through pleadings.

J. Vinelott further added that each event claimed must be identified separated and each of the events claimed qualifies the claimant to the benefit sought which in this instance is the composite sum claimed. This meant that each cause had resulted in the equivalent effect and these equivalent effects amount to the entire global sum claimed. Therefore, if any one cause were to succeed, there would be no need for apportionments of effects and the sums claimed. This suggestion of course made no reference to concurrent causes and their outcome. He stated the following as conditions to proceed with global claims:

> First, that a rolled-up award can only be made in a case where the loss or expense attributable to each head of claim cannot in reality be separated and secondly that a rolled-up award can only be made where apart from that practical impossibility the conditions which have to be satisfied before an award can be made have been satisfied in relation to each head of claim.

The acceptance of a global claim was thrown into question by *Wharf Properties Ltd and Another vEric Cumine Associates and Others.*[8] In this case the plaintiff made no attempt to link the cause with the effect in respect of a claim by the client against his architect or failure properly to manage, control, coordinate, supervise and administer the work of the contractors and subcontractors as a result of which the project was delayed. Six specific periods of delay were involved but the statement of claim did not show how they were caused by the defendant breaches. The plaintiff pleaded that due to the complexity of the project, the interrelationship and very large number of delaying and disruptive factors and their inevitable knock-on effects, it was impossible at the pleadings stage to identify and isolate individual delays in the manner the defendant required and that this would not be known until the trial.

In *Wharf Properties Ltd v Eric Cumine Associates*, the clients actions against their architects for negligent design and contract administration were struck out as incomplete and therefore disclosing no reasonable course of action and has been interpreted by some as a setback in the judicial approval of the global claims approach. As per Lord Oliver:

> The pleading is hopelessly embarrassing as it stands. In cases where the full extent of extra costs incurred through delay depend upon a complex interaction between the consequences of various events, so that it may be difficult to make an accurate apportionment of

---

[8] (1991) 52 BLR 1.

the total extra costs, it may be proper for an arbitrator to make individual financial awards in respect of claims which can conveniently be dealt with in isolation and a supplementary award in respect of the financial consequences of the remainder as a composite whole. This has, however, no bearing upon the obligation of a plaintiff to plead his case with such particularity as is sufficient to alert the opposite party to the case which is going to be made against him at the trial.

He further stated:

This claim is advanced not only without any specification of the causal connection between the breaches and the sums claimed but without any facts which will enable the defendant to ascertain which parts of these sums are being alleged to be attributable to the breaches alleged.

The immediate reaction to the decision of the Privy Council in *Wharf Properties v Eric Cumine Associates* was that it sounded the beginning of the end to global claims. As time passed, this view was correctly seen to be misplaced. *Wharf Properties* is properly analysed as a special case, determined on its own particular facts and procedural history. This case shows that those responsible for the preparation and presentation of global claims will need to work hard with those who have proper knowledge of the events so as to provide an adequate description of them. Equally, it will mean that proper records will need to be kept and as importantly good use will have to be made of existing records to provide the necessary details. It will no longer be possible to call in an expert who will simply list all the possible causes of complaint and try to avoid having to give details of the consequences of those events before proceeding.

While there have been some liberal views on global sums claimed, there has been less acceptance of rolled-up causes to rolled-up effects. Mr. Recorder Tackaberry QC, sitting as a deputy official referee in *Mid Glamorgan County Council v J Devonald Williams and Partner*,[9] highlighted the principles of justified global claims that the claimant must abide with: "(1) a proper cause of action must be pleaded; (2) that the specific events are relied upon, must not only be shown to satisfy the contractual requirement but also its causal effect; and (3) that the there must also be nexus between the event relied upon to the money claimed." However, he did agree that global claims were allowed where the extra costs claimed involved a complex interaction between various delaying events that it was impossible or impractical to plead specific causes to specific effects or specific money claims.

In the words of J. Byrne, in *John Holland v Kvaerner* (1996) (*supra*), global claims must be approached "with a great deal of caution" as it may be driven by a "desire to conceal its bogus nature by presenting it in a snowstorm of unrelated and insufficiently particularized allegations, or by a desire to disadvantage the defendant in some way." He further confirmed, *obiter dicta*, that the possibility of a court assisted claim, arguably based on a sense of perceived justice, may open the floodgates and encourage more unsubstantiated global claims with claimants

---

[9]   (1992) 8 Const. L.J. 61.

merely regurgitating all its available evidence requiring the respondent to stiff through the evidence to decipher the likely damages to various distinct causes and effects.

In *John Holland v Kvaerner*, the court also held that the eventual ability of the court to apportion damages to particular causes at the hearing would effectively mean that the claimant itself could have done so, and as such, shows that the global claim basis is unjustified. Alternatively, such an apportionment without evidence from the claimant would be akin to a guessing game or lottery. Flexibility to the strict requirements of pleadings does not justify the proliferation of evidence in such a broadly pleaded claim. In fact, the arguments against global claims have been centered on whether globally pleaded claims ought to be struck off even before the hearing stage.

The basic position is summarised, by the basic assumption, that contractors often have claims dependent on a number of separate causes, each of which has contributed to delay and extra cost. In principle, the loss attributable to each cause should separately be identified and particularised, but separation may be difficult. The law has developed as a careful balance between these practical difficulties for claimants and the rights of defendants to know the case against them in adequate detail. This leaves open the question of whether a global claim will succeed at trial. At trial, it should not be forgotten that you must prove your case and there is an important distinction between a global claim as a matter of pleading and a global claim as a matter of evidence. The position seems to be this:

(1) When a claim for an extension of time and/or loss and expense is advanced pursuant to contractual terms, then an arbitrator or the court can make a global award, subject to the same limitations as were set out in *Crosby* (1967) (*supra*) and *Merton* (1985) (*supra*). But the attitude is against merely impressionistic assessments, and a claimant is far more likely to succeed if he can show what effects flowed from what events giving rise to entitlement; and

(2) Where the claim is for damages for breach of contract, the claimant's task may be somewhat easier because he will usually be able to claim damages for losses, at any rate in the alternative, and under that head the arbitrator or the court is much more likely to be persuaded indeed is probably required to take an impressionistic approach.

Global claims at present ought to be well thought of before proceeding with the claim. While at the pleading stage some amount of flexibility can be allowed, it surely cannot be the same for the standard of proof. The limitations that global claims face can be summarised with the following points:

• Presently there are technological developments that if utilised will allow a fair and proper assessment of the dominant cause to the effect.
• A party cannot be allowed to analyse and provide evidence of a cause being a dominant cause and at the same time suggesting that all other causes are equally dominant.

- With critical path method techniques there can only be one dominant cause for any particular effect.
- The courts or the arbitrators should not be encouraged to try their hands at guess work apportionments while under the guise if performing justice. Justice must also be seen to be done.

In a global pleaded claim, the respondent will effectively have to reconstruct the progress of works and to analyse every step in order to establish possible defence without being able to focus on particular events occurring at particular times. Such a process required of the respondent would inevitably result in a longer arbitration period and higher costs of the arbitration. It entails studying voluminous documents discovered and obtaining extensive expert analysis on all possible cause and effects that did arise. Without the identification of the sums claimed for any particular effects, the respondent may not be able to concede to any sum claimed for any particular event and to resist the rest. This also hampers any form of culling down of the issues at the hearing and truly hampers early settlement of disputes.

6.4 Causation of Loss

Questions of causation in construction claims are best illustrated by global claims. The essence of the global claim is that there are a multitude of causes in terms of variations, breaches of contract and matters giving claims under the contract which can be proved on an individual basis. There is, then, an overall delay to the project and an overall increase in cost which exceeds the price. The global claim runs into difficulties where it can be shown either that some of the allegedly causative events do not lead to the delay or loss or that there are some causative events which are the responsibility of the party making the claim.

However, as pointed out in *John Doyle Construction Ltd v Laing Management (Scotland)* (2004) (*supra*), the draconian effect of total failure may be overcome by two mitigating factors. The first is that on the evidence there is an established causal connection which permits part of the claim to succeed. In this case, the obligation to plead a global case was described by Lord Young in these terms:

> In a case involving the causal links that may exist between events having contractual significance and losses suffered by the pursuer, it is obviously necessary that the events relied on should be set out comprehensively. It is also essential that the heads of loss should be set out comprehensively, although that can often best be achieved by a schedule that is separate from the pleadings themselves. So far as the causal links are concerned, however, there will usually be no need to do more than set out the general proposition that such links exist. Causation is largely a matter of inference, and each side in practice will put forward its own contentions as to what the appropriate inferences are. In commercial cases, at least, it is normal for those contentions to be based on expert reports, which should be lodged in process at a relatively early stage in the action. In these circumstances there is relatively little scope for one side to be taken by surprise at proof, and it will not normally be difficult for a defender to take a sufficiently definite view of causation to lodge a tender, if that is thought appropriate. What is not necessary is that averments of causation

should be over-elaborate, covering every possible combination of contractual events that might exist and the loss or losses that might be said to follow from such events.

The second is that causation must be treated as a matter of commonsense as stated by Lord Young:

> The second factor mitigating the rigour of the logic of global claims is that causation must be treated as a common sense matter. That is particularly important, in my view, where averments are made attributing, for example, the same period of delay to more than one cause.

In this connection, that the question of causation must be treated by the application of common sense to the logical principles of causation, it is frequently possible to say that an item of loss has been caused by a particular event notwithstanding that other events played a part in its occurrence. In such cases, if an event or events for which the client is responsible can be described as the dominant cause of an item of loss notwithstanding the existence of other causes that are to some degree at least concurrent that will be sufficient to establish liability.

Normally individual causal links must be demonstrated between each of the events for which the client is responsible and particular items of loss and expense. For a loss and expense claim under a construction contract to succeed, the contractor must aver and prove three matters; firstly, the existence of one or more events for which the client is responsible; secondly, the existence of loss and expense suffered by the contractor; and, thirdly, a causal link between the event or events and the loss and expense. As summarised by J. Humphrey LLoyd in *Bernhard's Rugby Landscapes Ltd v Stockley Park Consortium Ltd*[10]:

> A global claim in the sense used in argument is the antithesis of a claim where the causal nexus between the wrongful act or omission of the defendant and the loss of the plaintiff has been clearly and intelligibly pleaded. However that nexus needs not always be expressed since it may be inferred. There must be a discernable nexus between the wrong alleged and the consequent delay or money for otherwise there will be no agenda for the trial.

A party's success in advancing a global claim depends on its ability to prove that the other party was responsible for all the events which caused the global loss. The court, then, would be prepared to hear evidence so as to ascertain a causal link between the individual losses claimed and the individual events which the other party is alleged to have caused. In doing so, the court may ultimately award a sum which is less than the original global claim.

Frequently, however, the loss and expense results from delay and disruption caused by a number of different events, in such a way that it is impossible to separate out the consequences of each of those events. In that case, the events for which the client is responsible may interact with one another in such a way as to produce a cumulative effect. If, however, the contractor is able to demonstrate that all of the events on which he relies are in law the responsibility of the client, it is not necessary for him to demonstrate causal links between individual events and particular heads of loss. Therefore, because all of the causative events are matters

---

[10]   (1997) 82 BLR 39.

for which the client is responsible, any loss and expense that is caused by those events and no others must necessarily be the responsibility of the client.

A common example occurs when a contractor contends that delays and disruptions have resulted from a combination of late provision of drawings, specifications and design changes instructed on the client behalf; in such a case all of the matters relied on are the legal responsibilities of the client. Where, however, it appears that a significant cause of the delay and disruption has been a matter for which the client is not responsible; a claim presented in this manner can necessarily fail. If the claim is to fail, the matter for which the client is not responsible in law must play a significant part in the causation of the loss and expense. In some cases it may be possible to separate out the effects of matters for which the client is not responsible.

In *John Holland Construction & Engineering Pty Ltd v Kvaerner RJ Brown Pty Ltd* (1996) (*supra*), J. Byrne went on to consider the claim made by the plaintiffs in the case before him, and pointed out that, because it was a global claim, it was necessary to eliminate any causes of inadequacy in the tender price other than matters for which the client was responsible. It was also necessary to eliminate any causes of overrun in the construction cost other than matters for which the client was responsible. The logical consequence implicit is that the client breaches caused the extra cost or cost overrun. This implication is valid only so long as the client breaches represent the only causally significant factor responsible for the difference between the expected cost and the actual cost. He stated:

> It is the second aspect of the understated assumption, however, which is likely to cause the more obvious problem because it involves an allegation that the breaches of contract were the material cause of all of the contractor's cost overrun. This involves an assertion that, given that the breaches of contract caused some extra cost, they must have caused the whole of the extra cost because no other relevant cause was responsible for any part of it.

It is accordingly clear that if a global claim is to succeed, whether it is a total cost claim or not, the contractor must eliminate from the causes of his loss and expense all matters that are not the responsibility of the client. It may be possible to identify a causal link between particular events for which the client is responsible and individual items of loss. On occasion that may be possible where it can be established that a group of events for which the client is responsible are causally linked with a group of heads of loss, provided that the loss has no other significant cause.

In a case involving the causal links that may exist between events having contractual significance and losses suffered by the pursuer, it is obviously necessary that the events relied on should be set out comprehensively. It is also essential that the heads of loss should be set out comprehensively. So far as the causal links are concerned, however, there will usually be no need to do more than set out the general proposition that such links exist. Causation is largely a matter of inference, and each side in practice will put forward its own contentions as to what the appropriate inferences are.

The approach adopted by the courts can be pragmatic towards global claims. Therefore, a party wishing to prove causation in cases where there are complex facts has to ensure that, at both the pleading and proof stages, its case must be clear and easily understandable both by the other side and by the tribunal.

In the difficult area of proof of causation, properly prepared computer generated analyses such as critical path methods can be useful provided that they are founded in reality and are seen as a means to prove a case rather than proof in themselves. Whatever evidence is presented there must be a clear understanding of what it is meant to establish and the evidence must be in a form which can be readily understood by the tribunal. While a fundamentally poor case cannot be improved upon by the method of presenting evidence, there are many good arguable global claims where proper presentation of the evidence can assist in establishing the case.

6.5 Apportionment

In defending global loss and expense claims, clients will invariably argue that unless the contractor can show that all the events of delay and disruption and all the causes of the loss are the client responsibility, the global claim must fail. If the contractor is unable to prove that there was no concurrent cause of the delay and disruption for which the client was not responsible, it will be necessary for him to show that the causes of the delay and disruption for which the client is responsible are the significant or dominant causes. Contractors should, therefore, identify all the significant or dominant causes of the delay and disruption for which the client is responsible and which have caused them to incur loss and then make a reasonable attempt to allocate sums of loss to these causes or events. If the contractor fails to provide evidence of this, for instance in the form of daily or weekly site costs and their calculation and a detailed analysis showing when and how the significant or dominant causes made an impact upon the works, the adjudicator, court or arbitrator will be unable to make any apportionment of loss and the global claim will fail in its entirety.

The evidence presented at trial may allow specific breaches to be linked to specific losses. Further, it can be considered, as stated formerly, that causation is a matter of common sense and concurrent causes of financial loss may be determined by considering what the dominant cause of the loss was. Even if events could not be said to be the dominant cause of the loss, the court may still attempt to apportion the losses between the events that the defendant was responsible for and those for which it was not.

Thus, in *Lichter v Mellon Stuart Company*,[11] the plaintiffs total cost claim on one contract was rejected on the ground that a substantial amount of their loss was

---

[11] 305 F.2d. 216 (3d Cir. 1962).

the consequence of factors other than breaches of contract by the defendants. The court could find no basis for allocation of the plaintiff's claim, which was for a lump sum, between those causes which were actionable and those which were not, with the result that the entire claim was rejected. The Court of Claims had held that part of the plaintiff's extra cost on this contract was attributable to the fault of the defendant and part was attributable to other non-compensable factors. The Court of Appeals stated the result of that finding as follows:

> Once it had thus been established that only part of the claim represented extra cost chargeable to Mellon, the one question remaining was whether a reasonable allocation of part of the total sum was possible. The court undertook such an allocation, guided by evidence concerning the extra time required for the performance of the stone contract as the result of the improper shelf angles. We cannot say that this was an arbitrary method of allocation. Indeed, the plaintiff is not in position to complain that the allocation was imprecise since it bore the burden of proving how much of the extra cost resulted from Mellon's improper conduct. The plaintiff risked the loss of its entire claim, as occurred with reference to the masonry contract, if the court should not have been able to make a reasonable allocation.

The important points that emerge from this decision are, first, that the courts are willing to undertake an apportionment exercise and, secondly, that any such apportionment must be based on the evidence and carried out on a basis that is reasonable in all the circumstances.

This apportionment approach has been viewed by many commentators as a radical departure from the approach taken previously by courts but however it can still be possible for the courts to apportion losses as viewed in *John Doyle Construction Ltd v Laing Management (Scotland) Ltd* (2004) (*supra*), where Lord Young affirmed the above by stating:

> Even if it cannot be said that events for which the employer is responsible are the dominant cause of the loss, it may be possible to apportion the loss between the causes for which the employer is responsible and other causes. In such a case it is obviously necessary that the event or events for which the employer is responsible should be a material cause of the loss. Provided that condition is met, however, we are of opinion that apportionment of loss between the different causes is possible in an appropriate case. Where the consequence is delay as against disruption, that can be done fairly readily on the basis of the time during which each of the causes was operative. During the period when both operated, we are of opinion that each should normally be treated as contributing to the loss, with the result that the employer is responsible for only part of the delay during that period. Unless there are special reasons to the contrary, responsibility during that period should probably be divided on an equal basis, at least where the concurrent cause is not the contractor's responsibility. Where it is his responsibility, however, it may be appropriate to deny him any recovery for the period of delay during which he is in default.

In *John Doyle v Laing* (2004) (*supra*), and in relation to causation and apportionment, Lord Young opinion is clear as he stated:

> The fact that the pursuer has been driven (or chosen) to advance a global claim because of the difficulty of relating each causative event to an individual sum of loss or expense does not mean that after evidence has been led it will remain impossible to attribute individual sums of loss or expense to individual causative events. The point is illustrated in certain of

the American cases. The global claim may fail, but there may be in the evidence a
sufficient basis to find causal connections between individual losses and individual events,
or to make a rational apportionment of part of the global loss to the causative events for
which the defender has been held responsible.

Therefore, as stated in *John Doyle v Laing*, apportionment would be more
readily obtained where the loss was being calculated by reference to delay in the
works. Either the loss would be apportioned on the basis of the time during which
each of the causes was operative, or responsibility could be divided on an equal
basis. In carrying out such an apportionment, where a concurrent cause of delay is
the contractor responsibility it may be appropriate to deny him any recovery for
the period of delay during which the contractor is in default. This will make it all
the more important for contractors to avoid apportionment where they can by
demonstrating that the client events they rely upon are dominant. Matters become
more complex when considering disruption to the contractor work. Nevertheless,
apportionment will frequently also be possible in such cases and although that
might result in a somewhat rough and ready result, where the procedure is similar
to that that used in assessing contributory negligence.

The alternative to such an approach was a strict view that if a contractor
sustains a loss caused partly by events for which the client is responsible and partly
by other events; it cannot recover anything because it cannot demonstrate that the
whole of the loss is the responsibility of the client. However, the courts can regard
this as an unacceptable conclusion. The practical difficulties of carrying out an
exercise of apportionment should not prevent the contractor from being able to
recover at least some elements of a global claim.

In *City Inn Ltd v Shepherd Construction Ltd*,[12] in respect of loss and expense, it
was decided that the direct loss and expense and delay sustained by the contractor
could and should be apportioned between the events for which the client was
responsible and the events for which the contractor was responsible. In relation to
time and relevant event, Lord Young said this:

> While delay for which the contractor is responsible will not preclude an extension of time
> based on a relevant event, the critical question will frequently, perhaps usually, be how
> long an extension is justified by the relevant event. In practice the various causes of delay
> are likely to interact in a complex manner; shortages of labour will rarely be total; some
> work may be possible despite inclement weather; and the degree to which work is affected
> by each of these causes may vary from day to day.

6.6 A Case Law Approach

It is more than four decades since the courts opened the door to global claims,
recognising that a complex interaction between the consequences of the various
causes of loss might make it extremely difficult or even impossible to ascertain

---

[12] (2007) CSOH 190 and (2010) CSIH 68 CA 101/00.

with accuracy the effects of any single causative event. However, the door was only slightly ajar and, prior to admitting any global claim, the courts have systematically imposed strict conditions with respect to the pleadings and above all with respect to any contribution to the damage that may have been made by the claimant himself.

Since a global claim, in its crudest form, alleges that the entire difference between the contractor tender price and his actual costs has arisen from a number of breaches caused by the client, and makes little or no attempt to link the alleged effects to individual breaches and in the event that the contractor has himself made a material contribution to this increase in costs, either by under pricing his tender, by his own inefficiencies or otherwise, the client would be unduly penalised if the courts were to accept such a claim without imposing strict conditions.

The courts have, therefore, chosen to reject global claims where the client has been able to demonstrate that the contractor has made more than a trivial contribution to the alleged damage. This position has, unfortunately, often led to a result which is equally unjust. Not infrequently, it is the client who has obtained a windfall, a contractor claim that is justified to a large extent being rejected because of a single cause of damage attributable to the contractor himself for which it is impossible or impracticable to isolate the effects.

The recent cases have suggested that a court will generally not interfere with an allowance to a party of a global or total costs claim as long as the evidence supports the conclusion that the party has suffered a quantifiable loss. Some critics of global claims have described a global or a total costs claim as a technique to conceal claims lacking any real substance or degree of preparation and as embarrassing and an abuse of the process of the court and that judges should approach a global claim "with a great deal of caution, even distrust"[13] and that the court "should be assiduous in pressing the plaintiff to set out the nexus with sufficient particularity to enable the defendant to know exactly what is the case it is required to meet and to enable the defendant to direct its discovery and its attention generally to that case."[14]

However, these criticisms have been considered by the courts to be overstated. Accordingly, the courts will permit a global claim provided that it is impossible or impractical for a party to disentangle each part of the total damage attributable to each breach of the contract and recently the power of the court to strike out global claims has been very limited and only used where the claim is so evidently untenable that it would be a waste of resources for the global claim to be demonstrated only after a trial, or where the pleading is likely to prejudice, embarrass or delay the fair trial of the action. The question whether a pleading in any given case based upon a global claim, a total cost claim or some variant of this, is likely

---

[13] Wharf Properties v Eric Cumine Associates (1991) 52 BLR1.
[14] John Holland Construction & Engineering Pty Ltd v Kvaerner RJ Brown Pty Ltd (1996) 82 BLR 8.

to or may prejudice, embarrass or delay the fair trial of the action must depend upon an examination of the pleading itself and the claim which it makes.

Contractors who present global claims by way of justification of presenting claims in a global form usually quote the decision in the case of *J. Crosby and Sons Ltd v Portland Urban District Council* (1967) (*supra*). In this case the contract overran by 46 weeks. The arbitrator held that the contractor was entitled to compensation in respect of 31 weeks of the overall delay, and he awarded a lump sum rather than giving individual periods of delay against nine delaying matters. The respondent contested the award arguing that the arbitrator was wrong in providing a lump sum delay without giving individual periods in respect of each head of claim. In this case, it was held that where a claim depended on an extremely complex interaction in the consequences of various denials, suspensions and variations, it may well be difficult and even impossible to make an accurate apportionment of the total extra cost between the several causative events.

J. Donaldson, in *J. Crosby & Sons Ltd v Portland Urban District Council*, has positively advocated that global claims can be allowed after separating from the global claim the subject matters that can be claimed individually and submitting the rest of the conflicting as a separate global claim. In *Crosby*, J. Donaldson allowed the lump sum award on the basis of the arbitrator findings that it was impossible to assess the periods of delay and costs to each of the delaying events loss of productivity claimed. He stated:

> I can see no reason why the arbitrator should not recognise the realities of the situation and make individual awards in respect of those parts of individual items of the claim which can be dealt with in isolation and a supplementary award in respect of the remainder of these claims as a composite whole.

He added:

> The delay and disorganisation which ultimately resulted was cumulative and attributable to the combined effect of all these matters. It is therefore impracticable, if not impossible, to assess the additional expense caused by delay and disorganisation due to any one of these matters in isolation from the other matters.

Despite numerous cases concerning global claims since 1967, when the *Crosby* case tends to be relied upon in support of such claims, many of which have tended to deal with the structure of the pleadings rather than the nature of the financial assessment of the claim. The principle established, in the *Crosby* case, is however subject to a number of important qualifications. These include that the events which are the subject of the claim must be complex and must interact so that it is difficult, if not impossible, to make an accurate apportionment and there must be no duplication or unjust inclusion of profit in the issues and damages raised in the claim.

In *London Borough of Merton v Stanley Hugh Leach* (1985) (*supra*), a similar opinion has been ascertained by J. Vinelott:

> If application is made for reimbursement of direct loss and expense attributable to more than one head of claim and at the time when the loss or expense comes to be ascertained, it

is impractical to disentangle or disintegrate the part directly attributable to each head of claim then, provided of course that the contractor has not unreasonably delayed in making the claim and so has himself created the difficulty, the architect must ascertain the global loss.

He confirmed the principles and the directives as stated by J. Donaldson in *J. Crosby & Sons Ltd v Portland Urban District Council* (1967) (*supra*):

I need hardly say that I would be reluctant to differ from a judge of Donaldson J's experience in matters of this kind unless I was convinced that the question had not been fully argued before him or that he had overlooked some material provisions of the contract or some relevant authority. Far from being so convinced, I find his reasoning compelling. I think I should nonetheless say that it is implicit in the reasoning of Donaldson J.

The case of *British Airways Pension Trustees v Sir Robert McAlpine and Sons*[15] was to establish whether the courts would in turn be prepared to take a robust approach. There were defects in the work which were alleged to be due to faults by the architect, the contractor and others. The plaintiffs argued on a global basis by saying that the result of all the defect put together was a reduction in the value of the property in the sum of £3.1 m. The defendants requested that further and better particulars be provided in respect of the claim. They asked to be given detailed information as to how much of the loss in value could be attributed to each and every defect. As the plaintiff was not prepared or was unable to provide more detailed information an application was made to strike out the claim.

In *British Airways Pension Trustees v Sir Robert McAlpine and Sons*, J. Andrews ordered the claim to be struck out, but his decision was overturned by the Court of Appeal. Lord Justice Saville in summing up said:

The basic purpose of pleadings is to enable the opposing party to know what case is being made in sufficient detail to enable that party properly to answer it. To my mind it seems that in recent years there has been a tendency to forget this basic purpose and to seek particularisation even when it is not really required. This is not only costly in itself but is calculated to lead to delay and to interlocutory battles in which the parties and the courts pore over endless pages of pleadings to see whether or not some particular points have or have not been raised or answered when in truth each party knows perfectly well what case is made by the other and is able properly to prepare to deal with it. Pleadings are not a game to be played at the expense of citizens nor an end in themselves but a means to the end and that end is to give each party a fair hearing.

J. Andrews further stated:

This is again not a case in which it could be said that the plaintiff claims were fundamentally flawed, in the sense that no further particulars could assist their cause, nor a case where there had been an express refusal to provide further particulars or a contumelious disregard of court orders. The default of the plaintiffs, serious though it was, fell far short of calling for the draconian remedy of striking them out.

The advocation of the issue of global claims was emphasised in much detail by J. Byrne in *John Holland Construction & Engineering Pty Ltd v Kvaerner*

---

[15] (1994) 72 BLR 31.

RJ Brown Pty Ltd (1996) (*supra*). He defined the logic behind a global claim where a single sum is claimed which is the difference between the total actual cost and the contract price or valuation of the work. If the total claim is for more than one event, then it is a particular form of a global claim. He stated that:

> The claim as pleaded is a global claim, that is, the claimant does not seek to attribute any specific loss to a specific breach of contract, but is content to allege a composite loss as a result of all of the breaches alleged, or presumably as a result of such breaches as are ultimately proved. Such claim has been held to be permissible in the case where it is impractical to disentangle that part of the loss which is attributable to each head of claim, and this situation has not been brought about by delay or other conduct of the claimant. Further, this global claim is in fact a total cost claim. In its simplest manifestation a contractor, as the maker of such claim, alleges against a proprietor a number of breaches of contract and quantifies its global loss as the actual cost of the work less the expected cost. The logic of such a claim is this: (a) the contractor might reasonably have expected to perform the work for a particular sum, usually the contract price; (b) the proprietor committed breaches of contract;(c) the actual reasonable cost of the work was a sum greater than the expected cost. The logical consequence implicit in this is that the proprietor breaches caused that extra cost or cost overrun. This implication is valid only so long as, and to the extent that, the three propositions are proved and a further unstated one is accepted: the proprietor breaches represent the only causally significant factor responsible for the difference between the expected cost and the actual cost. In such a case the causal nexus is inferred rather than demonstrated. The understated assumption underlying the inference may be further analysed. What is involved here is two things: first, the breaches of contract caused some extra cost; secondly, the contractor cost overrun is this extra cost. The first aspect will often cause little difficulty but it should not, for this reason, be ignored. It is the second aspect of the understated assumption, however, which is likely to cause the more obvious problem because it involves an allegation that the breaches of contract were the material cause of all of the contractor's cost overrun. This involves an assertion that, given that the breaches of contract caused some extra cost, they must have caused the whole of the extra cost because no other relevant cause was responsible for any part of it.

J. Byrne went on to consider the claim made by the plaintiffs in the case before him, and pointed out that, because it was a total cost claim, it was necessary to eliminate any causes of inadequacy in the tender price other than matters for which the client was responsible. It was also necessary to eliminate any causes of overrun in the construction cost other than matters for which the client was responsible.

J. Byrne further commented that a claimant under a global claim must prove four elements to succeed:

- The claimant could reasonably have expected to perform the contract works within the labour hours allowed in its tender.
- The respondent breached the relevant terms of the contract causing the claimant to expend more labour hours than it allowed in its tender.
- The actual hours expended exceeded the tender allowance.
- The breaches of contract were the only "causally significant factor" explaining the labour hours overrun.

This approach from the Supreme Court of Victoria was adopted in English Law by the then J. Lloyd in *Bernhards Rugby Landscapes Ltd v Stockley Park*

Consortium Ltd (1997) (*supra*), where he also gave the plaintiff leave to amend the claim since "its current form is not so oppressive or abusive as to justify refusal of leave to amend. The deficiencies may be cured by the provision of particulars or in some other way."

In the case of *Bernhards Rugby v Stockley Park*, the plaintiff landscape contractor entered into an agreement under seal with Trust Securities Holdings for the construction of a golf course on a landfill site. The work was subject to delay and detailed and lengthy claims were submitted. It was alleged by the defendant that the claims were bound to fail due to a number of reasons one of which was that they contained global delay claims for variations. It was held by the court that the global claim was a total cost claim since the plaintiff had qualified its alleged loss by subcontracting the expected cost of the works from the final costs. The court held that such a claim was permissible if it was impractical to disentangle that part of the costs attributable to each head of claim and the situation had not been caused by the plaintiff conduct. Therefore, in such circumstances the inference was that the client breaches had led to additional costs and that the *causal nexus* was to be inferred rather than demonstrated.

According to J. Lloyd:

In this country the present rules do not envisage that, in general, a case will be judge managed. It is not for any court to direct a party as to the method by which its case should be established. Official Referees are expected to control the cases in their list and do so, either on application or, where permitted and appropriate, of their own motion, with a view to ensuring, within the rules of court, that the presentation of a case is such that it ensures that the issues raised by it are or will be clearly defined, as a matter of procedure, both with a view to trial and also to see that the parties should be aware of the strengths and weaknesses in their respective cases so that only those disputes which require to be tried should come to the court for decision. For present purposes the position may be restated as follows:-

1. While a party is entitled to present its case as it thinks fit and it is not to be directed as to the method by which it is to plead or prove its claim whether on liability or quantum, a defendant on the other hand is entitled to know the case that it has to meet.
2. With this in mind a court may—indeed must—in order to ensure fairness and observance of the principles of natural justice—require a party to spell out with sufficient particularity its case, and where its case depends upon the causal effect of an interaction of events, to spell out the nexus in an intelligible form. A party will not be entitled to prove at trial a case which it is unable to plead having been given a reasonable opportunity to do so, since the other party would be faced at the trial with a case which it also did not have a reasonable and sufficient opportunity to meet.
3. What is sufficient particularity is a matter of fact and degree in each case. A balance has to be struck between excessive particularity and basic information. The approach must also be cost effective. The information may already in the possession of a party or readily available to it so it may not be necessary to go into great detail.

In my judgment Schedule II is in reality either a total cost claim or in the nature of one since it appears to seek to recover all BRL's costs in the period after the original date for completion, even if its technique for doing so is to present those costs as notional rates etc. I therefore approach it with caution (but not yet with suspicion since there may be legitimate reasons why such a quantification is justifiable in the circumstances of this contract). The vice of this part of lies in the fact that BRL says that it is unable to apportion the overall costs or other figures to any of the variations or other events set out in schedule III. I have to say that I do not find this proposition one which is easy to accept. It is clear from the analysis which has been made by BRL for the purposes of its claims for extensions of time that it is possible to identify periods of delay for each of the principal events upon which reliance is placed. I do not see why it is not possible thereafter to spell out the nexus between those events which it is said caused the costs claimed in paragraph 57 to have been incurred and to sever the events which did not have that cause, e.g. because they were concurrent or did not cause any delay to completion.

The *Stockley Park* case represented the law on global claims until at least 2002 and it is characterised by careful balance that a party is entitled to present its case as it thinks fit but a defendant on the other hand is entitled to know the case it has to meet. What is sufficient particularity is a matter of fact and degree in each case. A balance has to be struck between excessive particularity and basic information. The approach must also be cost effective as cases can go on for a long time with great expenses. In *Bernhards Rugby v Stockley Park*, the court considered all the major cases concerning global claims and as a result has produced a good summary of the current position:

- While a court will approach a global claim or a total cost claim with caution, such claims may be the only way in which a plaintiff can establish its loss.
- A global claim is permissible where it is impractical to disentangle that part of the loss attributable to each head of claim, and the situation has not been brought about by delay or other conduct on the part of the plaintiff. In such circumstances the court infers that the defendant breaches caused the extra cost or cost overrun and the *causal nexus* was inferred rather than demonstrated.
- The power of the court to strike out is very limited and should only be used where the claim is so evidently untenable that it would be a waste of resources for this to be demonstrated only after a trial, or where the pleading is likely to prejudice, embarrass or delay the fair trial of the action.
- The question whether a pleading in any given case based upon a global claim, a total costs claim or some variant of this, is likely to or may prejudice, embarrass or delay the fair trial of the action must depend upon an examination of the pleading itself and the claim which it makes.
- The fundamental concern of the court is that the dispute between the parties should be determined expeditiously and economically and, above all, fairly, and while a plaintiff is entitled to present its claim as it thinks fit, on the other hand a defendant is entitled to know the case which it has to meet with as much certainty and particularity as is reasonable, having regard to the circumstances and to the nature of the acts themselves by which the damage is done.

In *John Doyle Construction Ltd v Laing Management (Scotland) Ltd* (2004) (*supra*), Lord MacFadyen went on to comment that global claims approach to "all or nothing" was mitigated by certain factors, which formed an important aspect of his decision. He confirmed, *obiter dicta*, that while the global claim may fail, it does not follow that no claim would succeed and that when a global claim is necessary because of the complexity of causation does not mean that, after evidence has been presented, it would be impossible to attribute individual sums of loss to individual events. Although the global claim may fail, there may still be in the evidence a sufficient basis to make some connection and make a rational apportionment of part of the global loss to the causes for which the other party is responsible.

Applying those principles to the *Doyle v Laing* case, Lord MacFadyen held that the case should proceed. He said that concurrent causes in which one cause was not the responsibility of the other party were best left for detailed consideration in subsequent proceedings. He also warned that it would be wrong to exclude the possibility that the evidence given could provide a satisfactory basis for an award of some lesser sum than the full global claim. On that basis it is difficult to see that a global claim would be rejected simply on the basis that it was global. Nevertheless, even in subsequent proceedings, a global claim would fail if a material part of the cause of the loss was an event for which the other party was not liable and if the evidence disclosed no rational basis for the award of any lesser sum.

Lord MacFadyen emphasised that allowing the case to proceed would not give an injured party a free hand to choose how to prove loss. If a lesser claim were to be made, then it must be done on the basis of evidence that was presented within the scope of the existing case as represented. This was a warning to those reluctant to attempt a full cause and effect analysis with the hope of proving its case in subsequent proceedings. In summary, Lord MacFadyen summarised the basics of global claims as follows:

- In a global claim the loss is attributed to the list of events without any specific link being made between each part of the loss and each event.
- A global claim for loss may be advanced and there was plenty of case law to confirm this.
- The logic of a global claim was that all the events that contributed to causing global loss must be the liability of the other party.
- If there are events for which the other party had no liability, the effect of upholding the global claim would be to impose a liability which, in part, was not justified. In such a situation, a global claim would fail. Advancing a global claim for loss, therefore, is a risky strategy.
- Failure to prove that a particular event was the liability of the other party would not be fatal to the claim if the remaining events for which the other party was liable were proved to have caused the global loss.

In *London Underground Ltd v Citylink Telecommunications Ltd*,[16] J. Ramsey highlighted that the essence of a global claim is that, while the breaches and the relief claimed are specified, the question of causation linking the breaches and the relief claimed is based substantially on inference, usually derived from factual and expert evidence. The consequence of undermining a global claim is that the claim will fail as a global claim.

---

[16] (2007) B.L.R.391.

Although the case deals with the review of an arbitrator award, *London Underground v Citylink* may be seen as an endorsement of the global claims principle, as set out in *Doyle v Laing* (2004) (*supra*), where it was suggested that where a global claim is presented and it can be shown that some of the events that caused loss were not the responsibility of the client, this does not necessarily mean that the entire claim should fail; it may be possible to apportion the loss between events that were the responsibility of the client and those that were not. This essentially makes it easier for a contractor to make a wide ranging global claim and hope that some of the evidence it presents in respect of the various elements of that claim can be determined by the court. While a claimant making a global claim runs the risk that the entire claim will be undermined, provided that sufficient convincing evidence is presented in respect of individual elements of a claim, those individual elements may succeed and a claimant may recover some of the losses it has claimed even though the claim does not succeed in its entirety.

J. Ramsey further stated, *obiter dicta*, that while a contractor may wish to adopt the global approach and throw every possible cause of loss at the client, hoping that the tribunal or court will deal with issues of causation and that at least some of its claim will succeed, such a speculative approach is no guarantee of success. Even a judicially endorsed global claim is no substitute for a well-argued, clearly-defined claim backed up by supporting evidence.

Reflecting on the various cases, the current position on global claims is that claimants are required to establish separately the causal link between each causal event and each amount of claim, although such separation may be difficult. The inability of the claimant to objectively identify each of the financial consequences of each and every event giving rise to the claim does not allow the responsible party to escape paying the damages. A global claim is likely to fail if the defendant events causing the alleged loss are shown not to be significant.

6.7   John Doyle: A New Approach

In John Doyle v Laing Management (2004) (*supra*), the court considered a number of related decisions made by courts in Scotland, England, the USA and Australia and established the approach the courts in Scotland will take when asked to consider a global claim. It is likely that all forms of tribunal in England (adjudicators, arbitrators and the courts) and elsewhere will adopt the approach taken by the Scottish Court of Session when dealing with global claims. Contractors everywhere should therefore follow the guidance given by the court upon the presentation of loss and/or expense claims and the evidence that should be provided to prove such claims.

In *John Doyle v Laing Management*, the defenders were the management contractors appointed to carry out the construction of a new corporate headquarters for the Scottish Widows Fund and Life Assurance Society, their client being a company known as Edinburgh Construction Services Limited. The project was divided into a number of distinct works packages. These included works packages known as WP2010 and WP2011. By an agreement dated 14 September and 28 November 1995 the pursuers were appointed works contractors in respect of WP2011, which consisted of certain works on the superstructure of the building. The global claim in this case relates to that works package.

The pursuers aver that they began work on WP2011 on 25 September 1995. According to the construction programme agreed between the parties, whereby the work on WP2011 should have been completed 28 weeks later, on 7 April 1996. In fact practical completion of WP2011 was achieved after 50 weeks, on 7 September 1996. The pursuers claim that they are entitled to an extension of time of 22 weeks for completion of WP2011, and to a revised completion date of 7 September 1996. They further sought decree ordaining the defenders to procure the ascertainment of the pursuers loss and expense incurred in consequence of delay and disruption in the completion of the contract work, and decree ordaining the defenders to procure the final adjustment of the contract sum.

The issues that were in dispute in the reclaiming motion related to the calculation of the loss and expense that is alleged to have been suffered by the pursuers in consequence of delay and disruption in the completion of the contract works, that delay and disruption having been caused, it is said, by events for which the defenders were responsible.

The second issue debated was the relevancy of the averments in support of the pursuers claim for loss and expense. The contention for the defenders was that these amounted to a global claim, that is to say, a claim in which the individual causal connections between the events giving rise to the claim and the items of loss and expense making up the claim are not specified, but the totality of the loss and expense is said to be a consequence of the totality of the events giving rise to the claim. The defenders submitted that the success of a global claim was perilled on the proposition that all of the causal factors were matters for which the defenders were legally responsible. If, therefore, one factor founded on as playing a material part in the causation of the global loss could be seen to be the responsibility of the pursuers, or at least not the responsibility of the defenders, a global claim could not be maintained. The Lord Ordinary, being Lord Young, held that the pursuer's averments of loss and expense were relevant, and the present reclaiming motion is against that part of his decision.

John Doyle loss and expense claim was made on a global basis due to its inability to identify causal links between each cause of delay and disruption and the individual consequences of these causes. The court had to consider how to deal with a claim where the causes of delay and loss included matters which were not the fault of the client; for instance, exceptionally adverse weather conditions or the contractor's own default. Unsurprisingly, and as it is common in cases of this type, Laing argued that because Doyle's claim had been presented upon a global basis and it was unfair for it to be responsible for loss caused by matters for which Laing was not responsible, the entire claim should be dismissed. It was common ground that in circumstances where the defendant was culpable for all the causes of loss and there was no need to apportion delay between specific events then a global claim in principle could be put forward. Laing contended, however, that where the defendant was not liable for all the causes, it followed that there was a need to apportion delay between specific events but which was precluded, by the format of the global claim, with effect that the claim must fail.

The court decided that, in principle, there is no problem with advancing a global claim if there are a large number of interacting events and the loss and expense attributable to each event cannot be shown. Provided the contractor can specify the events, the responsibility of the client for each of them, the client involvement in causing the global loss and the method by which the loss claimed has been calculated, the claim can succeed.

Lord Young began his discussion by pointing out that the case was not concerned with whether a global claim for loss and expense may relevantly be advanced by a contractor under a construction contract. The pursuers had averred that, despite their best efforts, it was not possible to identify causal links between each cause of delay and disruption and the cost consequences thereof. On that basis, the defenders accepted that the pursuers were in principle entitled to advance a global claim. The Lord Ordinary nevertheless reserved his opinion as to whether an averment of that nature was essential to the relevancy of a global claim, what is required to prove such an averment and what the consequences of failure to prove it might be.

Lord Young went on to analyse the nature of a global claim:

> Ordinarily, in order to make a relevant claim for contractual loss and expense under a construction contract (or a common law claim for damages) the pursuer must aver (1) the occurrence of an event for which the defender bears legal responsibility (2) that he has suffered loss or incurred expense, and (3) that the loss or expense was caused by the event. In some circumstances, relatively commonly in the context of construction contracts, a whole series of events occur which individually would form the basis of a claim for loss and expense. These events may inter-react with each other in very complex ways, so that it becomes very difficult, if not impossible, to identify what loss and expense each event has caused. The emergence of such a difficulty does not, however, absolve the pursuer from the need to aver and prove the causal connections between the events and the loss and expense. However, if all the events are events for which the defender is legally responsible, it is unnecessary to insist on proof of which loss has been caused by each event. In such circumstances, it will suffice for the pursuer to aver and prove that he has suffered a global loss to the causation of which each of the events for which the defenders is responsible has contributed. Thus far, provided the pursuer is able to give adequate specification of the events, of the basis of the defender's responsibility for each of them, of the fact of the defender's involvement in causing his global loss, and of the method of computation of that loss, there is no difficulty in principle in permitting a claim to be advanced in that way.

6.8   John Doyle: Apportionment Principle

In *Doyle v Laing*, the court agreed with the foregoing statements of the law made by J. Byrne in *John Holland Construction & Engineering Pty Ltd v Kvaerner RJ Brown Pty Ltd* (2006) (*supra*). In principle, the court deduced that it is accordingly clear that if a global claim is to succeed, whether it is a total cost claim or not, the contractor must eliminate from the causes of his loss and expense all matters that are not the responsibility of the client. Accordingly, this requirement is mitigated by certain considerations such as apportionment and concurrency.

Apportionment in this way, on a time basis, is relatively straightforward in cases that involve only delay. Where disruption to the contractor work is involved, matters become more complex. Nevertheless, apportionment will frequently be possible in such cases, according to the relative importance of the various caus-ative events in producing the loss. Whether it is possible will clearly depend on the assessment made by the judge or arbiter, who must of course approach it on a wholly objective basis. It may be said that such an approach produces a somewhat rough and ready result. Moreover, the alternative to such an approach is the strict view that, if a contractor sustains a loss caused partly by events for which the client is responsible and partly by other events, he cannot recover anything because he cannot demonstrate that the whole of the loss is the responsibility of the client. That would deny him a remedy even if the conduct of the client or the architect is plainly culpable, as where an architect fails to produce instructions despite repe-ated requests and indications that work is being delayed. In such cases the con-tractor should be able to recover for part of his loss and expense.

The *John Doyle Construction Ltd v Laing Management* case has not changed the law in the sense of formally overruling existing authority but there is definitely a change of emphasis in this case in favour of a more pragmatic approach to global claims by tribunals. One of the main issues it raised is that even if a global claim should fail; a lesser sum may be awarded based on the evidence placed before the court. This must inevitably create the perception of better prospects for global claims than before the decision.

John Doyle Construction Ltd v Laing Management signals at least three prin-cipal changes in emphasis each of which there is an encouragement to the pursuit of global claims:

1. The nature of the defence has altered. A client must now show that a cause of alleged delay and/or loss which was not his responsibility was significant. This is obviously a question of degree and highly fact sensitive.
2. There is an express recognition that apportionment of delay/loss might be possible, even if rough and ready results are generated.
3. The merits of a global claim and its prospects of success are unlikely to be determined at an interlocutory stage on an application to strike out.

However, the court agreed with the argument advanced by Laing that the logic of a global claim requires all the events that contributed to the global loss to be events for which the client is responsible. If they are not, an unjustifiable liability would be imposed upon the client if the global claim was allowed. Taking this logic one stage further, a global claim will fail if the client can show that he is not responsible for one or more of the events the contractor has alleged caused the global loss or there are other events that caused the loss, not advanced by the contractor, for which the client is not responsible.

The court did, however, give contractors hope of being able to recover at least some elements of a global claim as such: "Although a global claim might fail, this would not necessarily mean that no part of the claim would succeed."

The court also considered the issue of concurrent delay. If the court had accepted Laing argument and it had become necessary for contractors to prove that all the events which had caused the loss were attributable to the client, it would have been impossible for contractors to recover loss and expense where there are concurrent causes of delay where some are the fault of the client and some are the fault of the contractor.

In dealing with the issue of concurrent delay, the court drew a distinction between significant and insignificant causes of delay and disruption and confirmed that where there is a significant or dominant cause of delay and disruption for which the client is not responsible, the global claim will fail. Conversely, if there are two causes of delay and disruption but one is more significant or dominant than the other, then provided the contractor can show that the client is responsible for the dominant cause, the claim could still succeed.

Even in cases where events for which the client is responsible are not found to be the dominant cause of the loss, it may be possible on the evidence presented by the contractor to apportion the loss between causes for which the client is responsible and the other causes. Provided the contractor can show that the event for which the client is responsible is the significant or dominant cause of the loss, an apportionment of the loss between different causes should be possible.

On the issue of causation arising out of concurrent delays the court made it clear that a common sense approach must be applied, and depending upon the evidence placed before the court, there might be sufficient basis to find causal connections between individual losses and individual events, or to make a rational apportionment of part of the global loss to events for which the client was responsible. If the contractor is unable to prove that there was no concurrent cause of the delay and disruption for which the client was not responsible, it will be necessary for him to show that the causes of the delay and disruption for which the client is responsible are the significant or dominant causes.

Contractors should therefore identify all the significant or dominant causes of the delay and disruption for which the client is responsible and which have caused them to incur loss and then make a reasonable attempt to allocate sums of loss to these causes or events. If the contractor fails to provide evidence of this, for instance in the form of daily or weekly site costs and resources and a detailed analysis showing when and how the significant or dominant causes made an impact upon the works, the court or arbitrator will be unable to make any apportionment of loss and the global claim will fail in its entirety.

Lord Young added further clarification. As a prerequisite for a global claim to succeed, the claimant must eliminate from the causes of his loss and expense, all matters that are not the responsibility of the defendant. That position was however mitigated by three key considerations:

1. It may be possible to identify a causal link between specific events for which the client is responsible and particular items of loss. By such an approach parts of the claim are able to be extracted from the global claim and separately allocated to individual events. Individual causal links must normally be proved between each breach/claims event and each item of loss and expense. If this is impossible, the claims events can be pleaded as producing a cumulative effect as long as the contractor can show that all the events pleaded are the responsibility of the client.
2. If an event or events for which the defendant is responsible could be considered as the dominant or primary cause of an item of loss that would be sufficient to establish liability, notwithstanding the existence of other causes that are to some extent at least concurrent or secondary. However, even where the loss has been caused both by matters for which the client is responsible and by matters for which he is not responsible the claim can still succeed if those for which the client is responsible are the dominant cause of the loss.
3. Thirdly, even if it cannot be said that events for which the defendant is responsible are the dominant cause of the loss, it may be possible to apportion the loss between the causes for which the defendant is responsible and other causes. This apportionment would be more readily achieved where the loss was being calculated by reference to delay in the works for the loss could be apportioned on the basis of the time during which each of the causes was operative, or responsibility could be divided on an equal basis. Even where it is not possible to identify a dominant cause of the loss and the causes are truly concurrent a global claim may partially succeed. It may be possible for the tribunal to make an apportionment between those matters for which the client is responsible and those for which he is not responsible. In this way the tribunal could apportion liability for the loss and award the contractor a part of his global claim.

There are main changes in emphasis in the law in *John Doyle Construction Ltd v Laing Management* and each is, to a greater or lesser extent, encouraging of global claims. Whereas previously it was understood that any cause of loss shown to be not the responsibility of the defendant would be fatal to the global claim, it now appears that this only applies if the cause of loss is dominant and the court seemed comfortable with the idea of apportionment of loss by the tribunal between causes for which the client is and is not liable, even if this may be a rough and ready process.

The basic position can be summarised as if contractors often have claims dependent on a number of separate causes, each of which has contributed to delay and extra cost. In principle, the loss attributable to each cause should be separately identified and particularised. A pleading of global claim is only permissible when it is impossible to attribute a specific loss to a specific breach/event. In other words, if separation is difficult and unattainable then a global claim could be presented. It appears therefore that a global claim may still be acceptable, but only in situations where it is impractical to disentangle that part of the loss attributable to each head of claim and most importantly only in situation where the party making the claim has not caused some of the delay and additional expenditure to be incurred.

6.9 Causal Nexus and Factual Necessity

Flexibility to the strict requirements of pleadings does not justify the proliferation of evidence in such a broadly pleaded claim. In fact, the arguments against global claims have been centred on whether globally pleaded claims ought to be struck off even before the hearing stage. The courts have signalled a move away from striking out a global claim in the early stages of proceedings being brought. However, the grant of a second chance by rejecting an application to strike out does not necessarily mean that a global claim would be successful in the final analysis at trial. If the relevant events are sufficiently complex and inseparable and it can be demonstrated that there is no other reasonable alternative to a global claim format then the claim may survive but that does not necessarily mean it will succeed. The fragile nature of a global claim is another incentive to adopt good record keeping as standard practice; the more that can be proved by reference to records and documents, the more robust any claim will be.

It is claimed that the cost of particularising a complex claim may be so expensive and time consuming that it is disproportionate to the monies claimed. Central to the courts proceedings are the requirements of the overriding objective and proportionality and, therefore, to dismiss global claims is to ignore the experience of those assessing claims who are quite able to filter out the meritorious from the padding. In *GMTC Tools and Equipment v Yuasa Warwick Machinery*,[17] LJ Leggatt said that:

> The plaintiffs should be permitted to formulate their claims for damages as they wish, and not be forced into a straightjacket of the judge's or their opponents choosing. The fundamental concern of the court is that the dispute between the parties should be determined expeditiously and economically and, above all, fairly, and while a plaintiff is entitled to present its claim as it thinks fit, on the other hand a defendant is entitled to know the case which it has to meet with as much certainty and particularity as is reasonable, having regard to the circumstances and to the nature of the acts themselves by which the damage is done.

Likewise the courts have rejected global claims for proof of cause to effect but have on occasions allowed global quantum claims. These decisions are in line with notion that proof of liability and proof of quantum are two separate issues. If global claims are tolerated, it should be limited to quantum claims only but with a basis to calculate or extrapolate a daily cost to delays. It should not be taken as a

---

[17] (1994) 73 BLR 102.

dispensation of the standard of proof of cause to effect which is effectively the delay event and the period of delay caused thereby.

Alternatively, as seen in *London Borough of Merton v Stanley Hugh Leach* (1985) (*supra*), the other danger that arises is in the event that the composite quantum claimed has also contributed to the claimant's own delays and faults. Therefore, if a contractor has suffered a variety of delaying events, some caused by the client and some by himself, a global claim will not be acceptable.

The eventual ability of the court to apportion damages to particular causes at the hearing would effectively mean that the claimant itself could have done so, and as such, the global claim basis is unjustified. The possibility of such court assisted claim, arguably based on a sense of perceived justice, may open the floodgates and encourage more unsubstantiated global claims with claimants merely regurgitating all its available evidence requiring the respondent to stiff through the evidence to decipher the likely damages to various distinct causes and effects.

Alternatively, such an apportionment without evidence from the claimant would be akin to a guessing game or lottery. Further, the courts determination of the apportionment may never have been an issue ventilated and thus considered by the respondent who is caught off guard and thus deprived of natural justice. The courts or the arbitrators should not be encouraged to try their hands at guess work apportionments while under the guise of performing justice. Justice must also be seen to be done.

Although the position might be viewed differently today, it was often thought by those on the receiving end of a global claim that an application to strike it out should be made. In defending global loss and expense claims, clients will invariably argue that unless the contractor can show that all the events of delay and disruption and all the causes of the loss are the client responsibility, the global claim must fail. If the contractor is unable to prove that there was no concurrent cause of the delay and disruption for which the client was not responsible, it will be necessary for him to show that the causes of the delay and disruption for which the client is responsible are the significant or dominant causes.

Contractors should therefore identify all the significant or dominant causes of the delay and disruption for which the client is responsible and which have caused them to incur loss and then make a reasonable attempt to allocate sums of loss to these causes or events. If the contractor fails to provide evidence of this, for instance in the form of daily or weekly site costs and resources and their calculation and a detailed analysis showing when and how the significant or dominant causes made an impact upon the works, the court will be unable to make any apportionment of loss and the global claim will fail in its entirety.

A client might be well-advised to insist, by way of express contractual provisions that a contractor maintains the records recommended. This may limit the necessity and opportunity for making a global claim but it is unlikely it is suggested, to eliminate it entirely. If there exists a contractual obligation upon the contractor to maintain particular types of records, perhaps in a defined format with a certain degree of detail, then if a failure to do so results in the pursuit of a global claim, it is not difficult to see that this would be another factor for a tribunal to consider when determining the merits of a global claim.

References

Case Law

Bernhard's Rugby Landscapes Ltd v Stockley Park Consortium Ltd (1997) 82 BLR 39
British Airways Pension Trustees v Sir Robert McAlpine and Sons (1994) 72 BLR 31
City Inn v Shepherd Construction Ltd. (2007) CSOH 190
City Inn v Shepherd Construction Ltd. (2010) CSIH 68 CA 101/00
GMTC Tools and Equipment v Yuasa Warwick Machinery (1994) 73 BLR 102
Inserco Ltd. v Honeywell Control Systems (1998) EWCA Civ 222
J. Crosby & Sons Ltd. v Portland Urban District Council (1967) 5 BLR 121
John Doyle Construction Ltd. v Laing Management (Scotland) (2004) BLR 295
John Holland Construction & Engineering Pty Ltd. v Kvaerner RJ Brown Pty Ltd (1996) 82 BLR 8
London Borough of Merton v Stanley Hugh Leach (1985) 32BLR 51
London Underground Ltd. v Citylink Telecommunications Ltd. (2007) B.L.R.391
Lichter v Mellon Stuart Company 305 F.2d. 216 (3d Cir 1962)
Mid Glamorgan County Council v J Devonald Williams and Partner (1992) 8 Const. L.J. 61
Nauru Phosphate Royalties Trust v Matthew Hall Mechanical & Electrical Engineering Pty Ltd
        (1994) 2 V.R. 386
Wharf Properties v Eric Cumine Associates (1991) 52 BLR 1

Books

Beale H, Bishop W, Furmston M (2008) Contract: cases and materials. Oxford University Press,
        Oxford
Bramble B, Callahan M (2010) Construction delay claims. Aspen Publishers, Maryland
Callahan M (2010) Construction change order claims. Aspen Publishers, Maryland
Carnell N (2005) Causation and delay in construction disputes. Blackwell Publishing, Oxford
Davison P, Mullen J (2008) Evaluating contract claims. Wiley-Blackwell Publishing, Oxford
(2002) Delay and Disruption Protocol. Society of Construction Law, UK
Hart H, Honore T (1985) Causation in the law. Oxford University Press, Oxford
MacGregor H (2010) MacGregor on damages. Sweet & Maxwell, UK
Reese C (2010) Hudson's building & engineering contracts. Sweet & Maxwell,UK
White N (2008) Construction law for managers, architects, and engineers. Delmar Cengage
        Learning, USA

Chapter 7
Global Claims: Total Cost Methodology and Substantiation

7.1 Total Cost Methods: An American Law Approach

The expression global claim has normally been used in Scotland, England and other Commonwealth countries to denote a claim calculated in the foregoing manner. In the United States, the corresponding expression is the total cost claim. A total cost claim involves the contractor claiming that the whole of his additional costs in performing the contract has been the result of events for which the client is responsible. In relation to the remaining parts of the loss and expense claim, the contractor may seek to prove causation in a conventional manner.

This may be particularly useful in relation to the consequences of delay, as against disruption. The delay, by itself, will invariably have the consequence that the contractor site establishment must be maintained for a longer period than would otherwise be the case and frequently it has the consequence that engineers, foremen and other supervisory staff have to be engaged on the contract for longer periods. Costs of that nature can be attributed to delay alone, without regard to disruption. Moreover, because delay is calculated in terms of time alone, it is relatively straightforward to separate the effects of delay caused by matters for which the client is responsible and the effects of delay caused by other matters. For example, delay caused by late instructions and delay caused by bad weather can be measured in a straightforward fashion, subject only to the possibility that the two causes operate concurrently.

The conditions for the acceptance of total cost claims are more explicitly defined in the United States courts. These conditions identified include satisfying the following proof:

1. The contractor tender or estimate was reasonable.
2. The actual cost is fair and reasonable under the circumstances.
3. The contractor must establish that it was not responsible for any part of the increased cost.
4. There is no other practical method available to quantify the damages with reasonable degree of accuracy.

A. D. Haidar, *Global Claims in Construction*,
DOI: 10.1007/978-0-85729-730-3_7, © Springer-Verlag London Limited 2011

In *Lichter v Mellon*,[1] a subcontractor sued the prime contractor for the balance of a building subcontract for the balance due and breach of contract and the contractor filed a counterclaim. The subcontractor claimed that the breach occurred as a result of delays in the project, which resulted in the subcontractor being forced to speed up its work and perform inefficiently. The subcontractor did not itemise its damages and introduced testimony as to what it would have cost to perform all of the work if the undertaking had proceeded without untoward occurrences in the manner contemplated at the time of the contracting. It then introduced testimony as to the actual cost of the entire job as delayed, interrupted and hindered by all causes. On appeal, the court rejected the subcontractor total cost method, finding that:

> In these circumstances the subcontractor's inability to break down its lump sum proof of extra costs justifies the denial of any recovery if on the record any substantial part of the added cost of performance was chargeable to non-actionable causes rather than to a breach of contract by the contractor.

The Appellate Court affirmed the District Court findings, holding that:

> On the whole record, we think the court was justified in concluding that a substantial amount of the lump sum which the subcontractor proved as extra cost of the masonry work was a consequence of factors other than a breach or breaches of contract by the contractor. Since the court could find no basis for allocation of this lump sum between those causes which were actionable and those which were not, it was proper to reject the entire claim.

The courts must reconcile competing interests. While courts will not prevent a claimant from recovering for a delay or disruption merely because the claimant cannot precisely quantify its actual damages, they will not give the claimant a windfall by allowing it to shift all of its overruns to the owner. Courts that have recognised the total cost method have done so reluctantly. In *WRB Corp. v United States*,[2] the court stated the total cost method "has never been favored by the court and has been tolerated only when no other mode was available and when the reliability of supporting evidence was fully substantiated. The total cost method, although frequently asserted by contractors, should only be used under exceptional circumstances."

In *Phillips Construction Co Inc v United States*,[3] the plaintiff undertook the construction of a large housing project connected with an air force base. During construction, heavy rainfall and extensive flooding were encountered. Under the contract signed by the parties, the plaintiff assumed the risks incident to abnormal

---

[1] Lichter v Mellon Stuart Company 305 F.2d. 216 (3d Cir. 1962).

[2] 183 Ct. Cl. 409, 426 (1968).

[3] 394 F 2d 834 (1968). The Armed Services Board of Contract Appeals held extensive hearings on this contract dispute. Based upon substantial evidence, the board decided that the plaintiff-contractor had suffered a changed condition for which it was entitled to an equitable adjustment. The inability of the parties to agree amicably on the amount of the adjustment resulted in further hearings. Thereafter, the Board issued a second decision on quantum holding that the amount of plaintiff's equitable adjustment should be $89,000.

rainfall as such. Nevertheless, it claimed that its difficulties were greatly compounded by the inadequacy of the government designed drainage system for the project, and it sued for the loss that it said resulted from the defective drainage system. The Board of Contract Appeals, the body charged with determining the dispute at first instance, rejected a total cost claim by the plaintiff, because the plaintiff total loss was caused partly by matters for which the government were responsible and partly by the exceptional rainfall, for which neither party was responsible. Nevertheless, the board agreed with the plaintiff contention about the inadequacy of the drainage system and apportioned the plaintiff additional costs between flooding caused by defective drainage and other factors.

In *Phillips Construction Co Inc v United States,* the court had reservations about the application of the total cost method in presenting claims. J. Curiam opinion was as follows:

> This method of proving damage is by no means satisfactory, because, among other things, it assumes plaintiff's costs were reasonable and that plaintiff was not responsible for any increases in cost, and because it assumes plaintiff's bid was accurately computed, which is not always the case, by any means." However, the court did not refuse the total cost method outright: "Accordingly, the Board held that plaintiff was entitled to an equitable adjustment. However, since plaintiff had computed its claim on a total cost basis which necessarily included additional costs in both flooded and nonflooded areas, and since the Board was unable on the evidence before it to segregate costs properly applicable to each, it returned the case to the contracting officer for negotiations with plaintiff as to amount. In my judgment, this is not that extreme case where the total cost approach represents the only feasible method of computing the amount of an equitable adjustment due the contractor.

This exercise was upheld by the Court of Claims, which observed that "it represented the best judgment of the fact presented on the record before it, and this is all that the parties have any right to expect."

Boyajian v United States[4] is the landmark federal case rejecting the use of a total cost method for determining causes of delay and apportionment. In *Boyajian,* by applying the total cost method, the contractor sought damages for labour, overhead and material costs which were not covered by contract receipts even though these increases occurred during non-delay periods. The contractor calculated its damages by deducting both its anticipated and actual costs from the entire project amount under the total cost method, but did not itemise these damages. The court rejected the total cost method, finding that it was an unacceptable method for determining damages for breach of the contract. The court held that the contractor was barred from failing to differentiate between delay and non-delay periods and that it could not indiscriminately lump the damages together. In addition, the court found that the record was replete with production interruptions and delays that

---

[4] 423 F.2d. 1231 (U.S. Ct. Claims, 1970). The Boyajian case involved a contractor who sued the United States Air Force for breach of a contract for modulators and interval and dwell testers. The contractor claimed that it suffered delay damages as a result of testing procedures established by the Air force that were unreasonable.

were caused by events which were not attributable to the defendant, but for which the plaintiff made no adjustments whatsoever.

In *Boyajian v United States,* the sitting judge stated the following:

> In situations where the court has rejected the total cost method of proving damages, but where the record nevertheless contained reasonably satisfactory evidence of what the damages are, computed on an acceptable basis, the court has adopted such other evidence; or where such other evidence, although not satisfactory in and of itself upon which to base a judgment, has nevertheless been considered at least sufficient upon which to predicate a jury verdict award, it has rendered a judgment based on such a verdict.

The *Boyajian* case gave numerous reasons for rejecting the total cost method of recovery. It found that recovery of damages for breach of contract is generally not allowed unless acceptable evidence demonstrates that the damages claimed resulted from and were caused by the breach. Furthermore, the proper measure of damages is the amount of the plaintiff extra costs which are directly attributable to the defendant actions:

> This theory has never been favoured by the court and has been tolerated only when no other mode was available and when the reliability of the supporting evidence was fully substantiated. The acceptability of the method hinges on proof that (1) the nature of the particular losses make it impossible or highly impracticable to determine them with a reasonable degree of accuracy; (2) the plaintiff's bid or estimate was realistic; (3) its actual costs were reasonable; and (4) it was not responsible for the added expenses.

However, contrary to these basic *causal nexus* damage principles, no attempt is made in *Boyajian v United States* to relate any specific amount of increased costs to any particular alleged breach. Nor is any satisfactory explanation given as to why an attempt was not made or why it would not have produced reasonably accurate results. The court held that, based on the record, it was impossible to conclude that the plaintiff-contract loss, constituting the difference between the plaintiff-contract expenditures and its contract receipts, was reasonably to be equated with the increased costs directly resulting from defendant alleged breaches.

It is important to point out that, even though the court rejected the total cost method according to the factual circumstances in the *Boyajian* case and dismissed the contractor claims, it did not unilaterally reject such an approach altogether as long as there is reasonably satisfactory evidence of what the damages are, computed on an acceptable basis.

In *Huber, Hunt & Nichols, Inc. v Moore,*[5] the contractor claimed that the architect plans and specifications were negligently prepared and contained errors and omissions and that the architects were negligent and dilatory in approving change orders, approving shop drawings and in the overall supervision of the work. Relying on *Boyajian, (supra)* (1970), the court rejected the contractor claim that it should have been entitled to rely on the total cost theory stating:

---

[5] (1977) 67 Cal.App.3d 278.

It is obvious that contractor could have maintained a proper accounting system to establish its alleged damage proximately caused by defendant's alleged negligence, if it had desired to do so. If we were to accept Contractor's contention as the law of this state, the result would, for all practical purposes, nullify all laws regarding competitive bidding on public contracts. Under such a concept, contractors could submit any bid necessary to obtain the job knowing that the public agency (or its architects) would be required to pay whatever costs contractor incurred on the project if contractor could discover some error or omission however irrelevant in the plans and specifications. In the final analysis what Contractor actually complains of is that the amount of money which Owner paid Contractor under the 25 [change orders] and the time allowed for the changes or additional work was not sufficient to reimburse Contractor for its total cost and total delay. It was within Contractor's legal power to compute estimated change order costs in a manner which would compensate Contractor for its total loss. It failed to do so.

In *E. C. Ernst, Inc. v Koppers Company, Inc.,*[6] as work progressed, difficulties developed. Problems at the site required the defendant to modify the original plans for the job. These modifications, and other reasons, required the plaintiff to do extra work not contemplated in the original contract with the defendant. In addition, a variety of factors led to delays in completing the project and as a result, the plaintiff incurred extra expense. Both parties blamed the other for delays, engineering failure and inadequate supervision stemming from a purchase order for furnace construction at a steel mill. The court ruled in favour of the contractor, finding that though the contractor was responsible for all of the delays, the subcontractor failed to link the delays to its damages.

In *E. C. Ernst, Inc. v Koppers Company, Inc.*, the District Court denied damages for two reasons; firstly, it found that the plaintiff's method of proof was too hypothetical and artificial and secondly, the court rejected the total cost approach as a method of proving damages. On appeal, the Appelate Court, finding on the outset that the District Court incorrectly rejected the total cost method, because, damages need not be proved with mathematical certainty, only reasonable certainty. According to J. Seitz:

> The plaintiff sought to prove its damages by using a variation of the total cost approach. Essentially, this method requires calculation of actual cost and of cost under the contract. The contract figure is then subtracted from the actual cost to find damages due to delay. This is not to say that a plaintiff merely may label damages evidence as being under the total cost method and leave the matter at that. Under the total cost method, at a minimum the plaintiff must provide some reasonably accurate evidence of the various costs involved.

In *McDevitt & Street Co. v Department of General Services,*[7] a general contractor asserted a total cost claim for delay damages caused by an architect's errors and omissions. The contractor claimed that it was entitled to the difference between its original budget for labour costs, plus the amounts received in change orders for specific extra work, and its actual labour costs expended as a result of the delay. The hearing officer held that the owner was liable for the delay, but only

---

[6] F. Supp. 729 (WD Pa. 1979).

[7] Fla. 1st DCA (1979).

awarded the contractor's average daily cost times the number of days the project was delayed. The hearing officer held that the direct labour component of the contractor's claim was covered by previous change orders, "which had extended the project time and the contract sum." The contractor appealed, claiming that it was entitled to the total amount it expended over and above the sum of its contract price and the amounts it received in change orders.

In explaining the total cost method, the McDevitt Court quoted from the Court of Claims' opinion in *J.D. Hedin Construction v US*[8]:

> The exact amount of additional work which plaintiff had to perform as a result of the foundation problem is difficult, if not impossible, to determine because of the nature of the corrective work which was being performed. There is no precise formula by which these additional costs can be computed and segregated from those costs which plaintiff would have incurred if there had been no government-caused difficulties. However, the reasonableness and accuracy of plaintiff's estimate, which was prepared by an experienced engineer whose qualifications have been unchallenged, have been established. Defense counsel stated that the estimate was not challenged. The closeness of the bids gives support to reasonableness of the estimate. The bidders were three extremely experienced contractors of large construction projects. Plaintiff on prior occasions had successfully constructed a number of large projects for the Veterans Administration. Plaintiff has established the fact that it performed additional work. Moreover, the responsibility of defendant for these damages is clear.

In *Amelco Electric v City of Thousand Oak*,[9] as a result of the many changes and difficulty in reading the sketches, Amelco had to use staff with more experience than it had estimated. In addition, the work was carried out in a disorganised and unco-ordinated manner. Amelco had to increase its labour force and, like other contractors, was often required to delay or accelerate particular tasks and to shift workers among tasks to accommodate work by other trades. Losses were incurred due to decreases in productivity and efficiency. At trial the total global cost claim amounted to $2,224,842. The jury awarded Amelco compensatory damages of $2,134,586 plus prejudgment interest of $495,340 and costs of $134,841.33. City appealed on a number of grounds including failure to give notice of the claim as required by the contract and it failed the four part theory mainly the part where the contractor must establish that it was not responsible for any part of the increased cost.

In *Amelco Electric v City of Thousand Oak*, J. Brown stated:

> We conclude Amelco failed to adduce substantial evidence to warrant instructing the jury on the four part total cost theory of damages. In particular, Amelco failed to adduce evidence to satisfy at least the fourth element of the four part test, i.e., that it was not responsible for the added expenses. A corollary of this element of the test is that the contractor must demonstrate the defendant, and not anyone else, is responsible for the additional cost.

---

[8] J.D. Hedin Constr. Co. v. United States, 347 F.2d 235 (Ct. Cl. 1965a).

[9] (2000), 82 Cal.App.4th 373. Amelco contracted to carry out the electrical installation to a new Civic Arts Plaza on behalf of City for $6,158,378. During the construction process there were a large number of changes in design, every part of the electrical work being changed at least once.

As to the presentation of facts and causation, J. Brown further added in rejecting the total cost method:

> Here, as in Boyajian (1970) (supra), 423 F.2d 1231, Amelco alleged and the jury was instructed it could find a breach of the contract on numerous grounds, including breach of the implied warranty of correctness, breach of contract by preventing or hindering plaintiff's performance of the contract, providing an inadequate design, making excessive changes to the project, making changes in a disorganized manner, failing to properly coordinate the work of the multiple prime contractors, accelerating Amelco's work, and failing to make payments to Amelco in a timely manner. Amelco never attempted to demonstrate how a particular alleged breach caused certain damages. Rather, Amelco conceded no effort was made during the project to distinguish between those inefficiencies that were Amelco's and those believed to be the responsibility of the City (and presumably other prime contractors and subcontractors). Moreover, Amelco conceded it had been inefficient in performing the contract and that it had reduced its claim by an apparently arbitrary 5%.

The award was affirmed by the Court of Appeal which stated that the trial court had not erred in its instruction to the jury. The relevant instruction was that:

> "If you find that City breached or abandoned the contract, then Amelco is entitled to recover the reasonable value of the work performed by it less the payments made by the City, and less any costs incurred by Amelco which are not fairly attributable to the City." However, under these circumstances, "the jury should not have been instructed to calculate Amelco's loss from any breach of contract under a total cost measure of damages."

Assessment by the total loss method was not appropriate in *Amelco* because there was insufficient evidence to assess factors on an individual basis and even if there had been, a total loss award would have been appropriate due to the difficulties and inaccuracies inherent in recording, alternatively estimating many instances of reduced productivity or non-productive time which are factors which also existed in the *Crosby v Portland UDC*[10] and *LB Merton v Stanley*[11] cases.

7.2 Modified Total Cost Method

In the American cases before the Court of Claims, a further category is recognised that of a modified total cost claim. A modified total cost claim is more restrictive than the total cost method and involves the contractor dividing up his additional costs and only claiming that certain parts of those costs are the result of events that are the client responsibility. This terminology has the advantage of emphasising that the techniques involved in calculating a global total cost claim need not be applied to the whole of the contractor claim. Instead, the contractor can divide his loss and expense into discrete parts and use the total cost method for only one, or a limited number, of such parts.

---

[10] J. Crosby and Sons Ltd. *v* Portland Urban District Council (1967) 5 BLR 121.

[11] London Borough of Merton v Stanley Hugh Leach (1985) 32BLR 51.

The modified total cost method is similar to the total cost method except that in this approach the contractor bid estimate is adjusted to account for activities that were underbid or deemed to be his responsibility. The total cost differential is thus modified to eliminate cost factors that are the responsibility of the contractor and also correct inaccuracies in the original estimate. This makes the approach a more reliable method than the total cost method. In certain instances, the project is analysed retrospectively to determine what the project should have cost as a baseline instead of relying on the original estimate or its adjusted value.

The modified total cost method is the total cost method adjusted for any deficiencies in the contractor proof in satisfying the requirements of the total cost method. The total cost method is used as only a starting point with such adjustments thereafter made in such computations as allowances for various factors as to convince the court that the ultimate reduced figure fairly represented the increased costs the contractor directly suffered from the particular action of defendant which was the subject of the complaint.

In *Servidone Construction Corporation v the United States*,[12] Servidone Construction Corporation (Servidone) encountered unexpected site conditions when building an earthen dam for the United States Army Corps of Engineers (the Corps). In September 1981, the Corps awarded Servidone a contract to construct an embankment, a spillway, outlet works, and several roads on an earthen dam near the city of Dallas-Fort Worth. Servidone winning bid was $25,781,338.18.

Servidone began work in May 1982. Over the next two years, Servidone encountered many problems due to differing site conditions. After incurring costs well beyond its bid price, Servidone filed a certified claim for equitable adjustment of the contract on March 1, 1984. Servidone did not complete the contract work until August 1985. On June 4, 1984, Servidone filed suit under the Contract Disputes Act, in the Claims Court. Servidone complained that the Corps breached an implied duty to provide adequate information for contract performance. Servidone also complained that it encountered unusual soil conditions covered by the contract's differing site condition. Finally, Servidone complained that the Corps caused delays by excessive quality assurance testing.

The court found the government liable for Servidone increased costs and computed damages with the total cost method. The court determined that Servidone incurred $23,703,582.00 in costs above its estimated costs or bid. However, the court also found that Servidone bid was too low. To compensate for Servidone unreasonably low bid, the trial court substituted a reasonable bid in the damages computation. This substitution reduced Servidone claimed costs by $9,262,459.00. These findings produced an award to Servidone of $14,441,123.00. The trial court awarded Servidone interest on this sum from the date the government contracting officer received Servidone certified claim.

The court held, *obiter dicta*, that to receive an equitable adjustment from the government, a contractor must show three necessary elements; liability, causation

---

[12]   931 F.2d 860. April 24, (1991).

and resultant injury. The court further held that to show the amount of injury, the contractor must show:

(1) The impracticability of proving actual losses directly.
(2) The reasonableness of its bid.
(3) The reasonableness of its actual costs.
(4) Lack of responsibility for the added costs.

Although finding Servidone bid unreasonable, the court awarded damages. In doing so, the court employed a modified total cost method. This modified method substituted a reasonable bid amount for Servidone original estimate.

The main points that can be extracted from this case are that a trial court must use the total cost method with caution and as a last resort. Under this method, bidding inaccuracies can unjustifiably reduce the contractor estimated costs. Moreover, performance inefficiencies can inflate a contractor costs. These inaccuracies and inefficiencies can thus skew accurate computation of damages. Despite this risk, this court predecessor condoned the total cost method in those extraordinary circumstances where no other way to compute damages was feasible and where the trial court employed proper safeguards.

The court found that Servidone met the four part test and thus approved the total cost method in this case. The court granted Servidone a recovery under the modified total cost method:

> The total cost approach was used as only a starting point with such adjustments thereafter made in such computations as allowances for various factors as to convince the court that the ultimate, reduced, figure fairly represented the increased costs the contractor directly suffered from the particular action of defendant which was the subject of the complaint.

In *Biemann and Rowell Co. v Donohoe Companies Inc.*,[13] a ventilating subcontractor sued the general contractor for breach of contract in the construction of a hospital at the University of North Carolina. The ventilating subcontractor used a total cost method of calculating damages. The court rejected this method on the basis that the total cost method is condoned only where no other way to compute damages is feasible, because it assumes that every penny of the plaintiff costs are *prima facie* reasonable, that the bid was accurately and reasonably computed and that the plaintiff is not responsible for any increases in cost.

The Superior Court found that Biemann failed to establish the causation element by proving that the delays by the general contractor caused the ventilating subcontractor delays. On appeal, the North Carolina Court of Appeals affirmed. The Appeals Court reasoned that it is well settled that a plaintiff has an obligation to prove the facts that will create a good basis for the calculation of damages. For the breach of an executory contract, a plaintiff may recover only such damages as can be ascertained and measured with reasonable certainty. Where both parties contribute to the delay, neither can recover damages, unless there is proof of clear apportionment of the delay and expense attributable to each party.

---

[13]   556 S.E.2d 1, 5 (N.C. Ct. App. 2001).

In *Biemann v Donohoe*, the court applied a four part test for recovery under the modified total cost method articulated in *Servidone Construction Cororporation v United States* (1991) (*supra*), and *Boyajian* (1970) (*supra*), which are: "(1) the impracticability of proving actual losses directly; (2) the reasonableness of its bid; (3) the reasonableness of its actual costs; and (4) the lack of responsibility for the added costs." The court held, *obiter dicta*, that the modified total cost method is the total cost method with adjustments for any deficiencies in plaintiff's proof in satisfying the four requirements. The modified approach assumes the elements of a total cost claim have been established, but permits the court to modify the test so that the amount plaintiff would have received under the total cost method is only the starting point from which the court will adjust the amount downwards to reflect the plaintiff's inability to satisfy the test.

In *Biemann v Donohoe*, the court held that Biemann and Rowell has failed to establish the elements of a total cost method claim as such:

First, Biemann and Rowell has not shown the Court that proving direct actual losses was impracticable. To establish the impracticability of proving actual losses directly, a plaintiff must show that the nature of its losses makes it impracticable to determine the amount of the actual losses with a reasonable degree of certainty. Biemann and Rowell failed to convince the Court that it was unable to assess direct losses attributable to Donohoe. Testimony by Biemann and Rowell employees indicated that Biemann and Rowell's accounting department kept records of labor overrun throughout the Neuropsych project. Such records could have been tied to instances where Donohoe was responsible for delays. In fact, Biemann and Rowell did not produce any accounting records, nor did they offer as witnesses any accounting personnel. Biemann and Rowell could have maintained records of delay costs, but did not do so. Second, Biemann and Rowell failed to prove that its bid on the Neuropsych project was reasonable. A determination of whether a bid was reasonable is made from the bids of others. The record contains no evidence of other bids submitted for the HVAC work on the Neuropsych project. Accordingly, the Court has no means by which to assess whether Biemann and Rowell's bid was reasonable. Biemann and Rowell's own employees testified that the bid was aggressive. Third, Biemann and Rowell failed to adequately allocate responsibility for its extra costs. "It is incumbent upon plaintiff to show the nature and extent of the various delays for which damages are claimed and to connect them to some act of commission or omission on defendant's part." 27 Fed. Cl. at 546 quoting Wunderlich Contracting Co. v United States, 174 Ct. Cl. 180, 200 (1965b). Biemann and Rowell's expert allocated responsibility for a narrow set of costs to Biemann and Rowell and attributed the remainder of the cost overrun entirely to Donohoe. Biemann and Rowell made no attempt to isolate the nature and extent of specific delays but rather attributed the entire project delay to Donohoe's failure to provide a temporary seal. In fact, the Court has found above that Biemann and Rowell also contributed to the overall project delay. Where both parties contribute to the delay, neither can recover damages unless there is proof of clear apportionment of the delay and expense attributable to each party. In conclusion, the total cost claim methodology used by Biemann and Rowell fails to establish damages to a reasonable certainty. Biemann and Rowell failed to verify the validity and accuracy of its data used in the calculations of damages. Specifically, Biemann and Rowell failed to account for factors such as: (1) the design, (2) acts of other co-primes and their subcontractors, (3) acts of the Owner, (4) acts of Biemann and Rowell and its subcontractors, (5) omissions and errors in Biemann and Rowell's bid, (6) mathematical computation errors and (7) the use of the Eichleay formula, which is appropriate for suspension of work and but not delay damages. These failures render Biemann and Rowell's calculations and assessment of damages unreliable and speculative.

In *Propellex Corporation v Brownlee*,[14] Propellex requested an equitable adjustment of the contract price, asserting that faulty government testing caused it to incur additional costs and filed a claim with the contracting officer in the amount of $1,790,065 for both contracts. The contracting officer issued a final decision admitting some culpability and allowing recovery of $77,325. Propellex appealed the contracting officer's final decision to the Armed Services Board of Contract Appeals (Board). Propellex presented its case for damages before the Board using a modified total cost method. In furtherance of its modified total cost claim, Propellex contended that it was impracticable to prove its claimed losses directly.

In its *Propellex Corpration v Brownlee* decision, the Board clarified the requirements for recovery of damages in government contract disputes using a modified total cost method on the basis that Propellex had not established the impracticability of proving its actual losses directly. The Board decision clarified the first of the four *Servidone* (1970) (*supra*) proof prerequisites which requires that, in order to recover damages under the total cost method, a contractor must first establish the impracticability of proving its actual losses directly. Here, the Board held that a contractor cannot establish the impracticability of proving its actual losses directly by unreasonably failing to keep records of its actual costs.

The Board held that using actual cost data to establish the amount of an equitable adjustment for additional work is the preferred method of proof as actual cost data as it "provides the court, or contracting officer, with documented underlying expenses, ensuring that the final amount of the equitable adjustment will be just that–equitable–and not a windfall for either the government or the contractor." The court also held that in the absence of actual cost data, contractors may use estimates to establish the amount of an equitable adjustment for additional work. The Board, however, noted in its decision that:

> Under its modified total cost method claim, Propellex still had the burden of proving the four requirements for a total cost recovery set forth above. The modified method simply was a way of easing that burden somewhat.

The Board determined, however, that Propellex had not established the impracticability of proving its added costs directly (total cost method requirement one) or that it was not responsible for the added costs (total cost method requirement four).

On appeal, Propellex challenged both of these rulings. First, it argued that the Board erred in concluding that it had not established the impracticability of proving its actual losses directly. Second, it contends that the Board erred when it concluded that the record did not allow for the removal of certain costs from Propellex's

---

[14] 342 F.3d 1335 (2001). The U.S. Army Armament, Munitions and Chemical Command awarded Propellex two firm fixed-price contracts to deliver Mark 45 electric gun primers to the U.S. Navy for a combined total price of approximately $2.6 million. The Army determined that lot six under the first contract did not meet contract requirements because black powder samples exceeded the maximum allowable moisture content limit. When Propellex completed this investigation, it informed the Army that it found no evidence to indicate that the moisture content of its black powder was excessive. The Army ultimately accepted all of the primers that Propellex produced.

modified total cost claim. The Appellate Court held that because substantial evidence supports the Board's conclusion that Propellex did not establish the impracticability of proving its actual losses directly and because the Board's decision is otherwise free of legal error, the Appellate Court affirms the decision of the Board and therefore under these circumstances, it is not necessary to decide the issue of whether Propellex satisfied the fourth requirement of the total cost method.

In *Propellex Corpration v Brownlee,* J. Schall stated the following:

Where it is impractical for a contractor to prove its actual costs because it failed to keep accurate records, when such records could have been kept, and where the contractor does not provide a legitimate reason for its failure to keep the records, the total cost method of recovery is not available to the contractor. In sum, substantial evidence supports the Board's conclusion that Propellex did not establish the impracticability of proving its actual losses directly and thus cannot recover under the modified total cost method. We do not agree with Propellex that the Board erred as a matter of law when it contrasted Propellex's ability to estimate certain costs that were not related to the moisture investigation with its inability to directly prove the investigation's costs. Contrary to Propellex's view, under the Board's ruling, compliance with requirement four of the total cost method, i.e., removing the costs that were not related to the moisture investigation, does not make it impossible to establish the first requirement, i.e., the impracticability of proving the contractor's losses directly. That is so because a contractor can always show why a court should not rely on its ability to segregate and remove certain costs in determining whether the contractor established the first requirement. The four requirements of the total cost method are distinct requirements and a contractor must prove all of them before it can obtain the benefit of the total cost method. Accordingly, it was not error for the Board to rely on Propellex's ability to estimate the costs that were not related to the moisture investigation as undercutting Propellex's argument that it is impracticable for it to prove its losses directly.

J. Schall further stated:

We do not agree with Propellex's analysis of Servidone. The use of the modified total cost method was appropriate in Servidone, not because of the nature of the contract work, but because that was the only meaningful way to express the difference between the contractor's reasonable anticipated costs and its reasonable actual costs. In other words, in Servidone, there was no way for the contractor to segregate the costs for the additional work it had to perform to complete the project. To the contrary, in this case, Propellex could have measured the additional investigative costs related to the moisture problem by setting up its accounting system to measure such costs. Accordingly, unlike the contractor in Servidone, Propellex is not entitled to recover under a modified total cost method.

In *Dillingham-Ray Wilson v City of Los Angeles,*[15] a recent California Court of Appeal case unequivocally established the modified total cost theory of damages

---

[15] (2010) 182 Cal.App.4th 1396. The City of Los Angeles initiated a competitive bidding process for work at the Hyperion Wastewater Treatment Plant. Contractor Dillingham-Ray Wilson was awarded the public works contract. Once construction began, the City issued over 300 change orders containing more than 1,000 changes to the plans and specifications. City requested an estimate of the cost of work and directed the contractor to begin work, stating that the parties would negotiate a lump sum payment at a later date. However, upon project completion, the City refused to pay the contractor a lump sum for the outstanding change order work, assessed liquidated damages and refused to release retention funds from escrow.

as a viable remedy for contractors. The court held, *obiter dicta*, that where no contractual requirement exists for a contractor to document its actual costs, that contractor may be able to recover for work it performed in good faith using the modified total costs method.

In this case, the Appellate Court disagreed with the trial court and found that the modified total cost theory was recognised as valid in *Amelco Electric v City of Thousand Oak* (2000) (*supra*) and that Dillingham could potentially recover under this theory, depending on interpretation of the contractual requirements for documentation of the actual costs of change orders. The court also reconciled with the common law requirement that only the best evidence of damages, not exact proof, and that the '*measure of damages*', but not the '*method of proof*' of those damages is required, and that therefore, Dillingham could potentially use engineering estimates to prove its damages.

In *Dillingham-Ray Wilson v City of Los Angeles*, the court remanded the issue of whether Dillingham contract required it to document its actual costs on the change orders issued by the City to the trial court for further proceedings. The court held, *obiter dicta*, that if Dillingham was not contractually required to document its actual costs, which Dillingham contended was not possible then he could use engineering estimates to prove its claims under a modified total cost method, provided those estimates were the best evidence available. The judge stated the following:

> Because Amelco recognizes that a contractor can recover on a modified total cost theory, that remedy is available in California. The trial court abused its discretion by not following the confirmed law set forth in Amelco and by declining to decide whether DRW demonstrated a prima facie case for determining damages based on a modified total cost theory. On remand, DRW may pursue a modified total cost theory of proving damages if DRW is not required to document its actual costs. If the trial court finds a prima facie case, then DRW shall be entitled to present a modified total cost theory to the jury.

7.3 Total Cost Method Essential Criteria

The First Criteria—No other practicable means of measuring damages—The first criterion for using the total cost method is that there are no other practicable means of measuring damages. The claimant must prove that its additional costs (resulting from the owner's actions) cannot be measured with any reasonable accuracy. In *J.D. Hedin* (1965a) (*supra*), for example, the court found that the exact amount of additional work the plaintiff had to perform was difficult, if not impossible, to determine. The court noted that the plaintiff had established that there was no precise formula by which its additional costs could be '*computed and segregated*' from those costs which it would have incurred if there had been no government caused difficulties.

A claimant can also meet this criterion by demonstrating that the defendant caused so many delays, disruptions or other changes that it was impossible to quantify the damages for each particular act.[16] Courts have reasoned that it is impractical for a claimant to maintain detailed cost records to measure the precise impact of such changes. However, a claimant will not satisfy this criterion by simply failing to maintain adequate records.

In *Boyajian* (1970) (*supra*), the court held that the contractor's failure to record its increased costs did not necessarily mean that there was no other method for calculating damages. The court recognised that it is unusual for contractors to keep detailed costs records of these impacts. Such failure, however, normally does not prevent the submission of reasonably satisfactory proof of increased costs incurred during certain contract periods or flowing from certain events based, for instance, on acceptable cost allocation principles or on expert claims consultants by showing that it was able to achieve estimated productivity rates during unimpacted periods of the project. However, this approach requires the claimant to establish an un-impacted period of comparable work. In short, a contractor must attempt to make a causal connection between the defendant's breach and its damages or explain why such an attempt was not made or why it would not have produced reasonably accurate results.

*The Second Criteria—The original bid or estimate was realistic—*The second criterion for using the total cost method is to establish that the claimant's bid or estimate is realistic. To satisfy this criterion, courts have typically required a claimant to show that it was diligent in preparing its bid and that the bid was within the range of other bids submitted for the project.

In determining that a claimant's bid was reasonable, these courts relied on testimony concerning the bidders' qualifications, the plaintiff's estimates for other similar projects and the methods used and information relied on in preparing the bid. Industry estimating manuals and comparisons of supplier and subcontractor quotes with bid amounts and material quantity estimates can also be used to support this testimony. It is important to note, however, that this type of an analysis is very expensive and time consuming; it is also easily refutable because, like the total cost method, it relies on assumptions, which claimants make during the bidding process.

*The Third Criteria—Actual costs were reasonable—*The third criterion for using the total cost method is that the claimant's actual costs were reasonable. The reasonableness of a claimant's actual costs generally is the easiest of the criteria to establish. Courts typically require a contractor to demonstrate that it acted reasonably in incurring its additional costs.

Contractors generally will attempt to satisfy this requirement through the use of expert testimony and reliance on industry standards. A contractor can also satisfy this criterion by demonstrating that it took measures to mitigate its additional costs.

*The Fourth Criteria—The contractor was not responsible for added expenses—*The final, and most difficult, criterion for using the total cost method is that the

---

[16] E. C. Ernst, Inc. v Koppers Company, Inc. 476 F. Supp. 729 (WD Pa. 1979).

claimant was not responsible for the additional costs incurred. This criterion is based on the contractual principle that damages are awarded only for costs incurred as a result of defendant's breach.

This criterion is particularly difficult to satisfy in complex construction cases in which both owner and contractor usually are responsible for delays and disruptions. In these cases, courts will not award total costs because such an award would compensate a claimant for its own errors and omissions.[17] Courts will consider several factors in determining whether a contractor is responsible for the added costs, including the claimant's performance and its experience with a particular type of project.

7.4 Cardinal Change

The cardinal change doctrine developed in courts as a means for contractors to avoid contractual limitations on damages in situations where changes grossly exceed the scope of the original contract. As well, when the claimant can prove that a change is a cardinal change, he can recuperate his costs easier using a total cost method. There is no easy formula which can be used to determine whether a change is beyond the scope of the contract and, therefore, the client is in breach of it. Each case must be analysed on its own facts and in light of its own circumstances, giving just consideration to the magnitude and quality of the changes ordered and their cumulative effect upon the project as a whole.

In making a determination, if a change is a cardinal change, the court will look at all relevant circumstances including but certainly not limited to the increase in cost of completing the contract and the number of changes made. Conversely, there is a cardinal change if the ordered deviations altered the nature of the thing to be constructed. In *General Contracting and Construction Co. v U.S.*,[18] the change was the deletion of one building from a hospital complex. The deletion represented 10% of the cost of the overall work. The court found that deleting an entire building was a fundamental change in the character of the project which the contractor had contracted to build. The court found a cardinal change not because of the magnitude of the change, but because of the quality of the change.

In *P. L. Saddler v U.S.*,[19] the contract was for the construction of a levy. The government modified the contract so that the length of the levy increased by 100%, from 1,000 to 2,000 feet, and the volume of the levy increased by 141%, from 5,500 cubic yards to 13,264 cubic yards. Converse to general contracting, the court found a cardinal change not because of the quality of the change but because of the magnitude.

---

[17] G.M. Shupe, Inc. v. United States, 5 Cl. Ct. 662 (1984).

[18] 84 Ct.Cl. 570 (1937).

[19] 287 F.2d 411 (Ct.Cl. 1961).

In *Luria Brothers and Company, Inc. v U.S.*,[20] the contract was for the construction of an airplane hangar. During the course of construction, the client completely changed the design of the foundation, which was the major structural component of the building. The contractor had to tear out previously completed work. When the contractor began excavating the foundation pursuant to the redesigned plans, the client abandoned those plans and required the contractor to excavate the foundation on a trial and error basis. The contractor had to dig the foundation foot by foot, stopping each foot so the client could test the stability of the soil, until the client was satisfied. The contractor ultimately had to excavate the foundation to elevations well below those on the amended plans. The original 330 day contract was extended an additional 518 days, a 160% overrun. The contractor claimed cost overrun was $248,665.76 on a $1,700,166.50 contract.

In *Luria Brothers and Company, Inc. v U.S.*, the court found a cardinal change not because of the magnitude of the overruns in time and money, but because in the change of the quality of the project. The court viewed the foundation as a major structural component of the project. Requiring the contractor to tear out its extensive completed work, materially amending the plans, abandoning the plans altogether in favour of an enormously labour and time intensive trial and error method and ultimately requiring excavation to depths much deeper than those contracted for, all combined to create a change of such magnitude that it breached the contract. In this case, the court found a cardinal change because of the qualitative changes in the contractor work, not because of the magnitude of time or cost overruns.

In *Allied Materials & Equipment Co. v United States*,[21] the purpose of the cardinal change doctrine was defined as "to provide a breach remedy for contractors who are directed by the Government to perform work which is not within the general scope of the contract, in other words, work which fundamentally alters the contractual undertaking of the contractor."

The best overall summary of the doctrine is in *Atlantic Dry Dock Corp. v U.S.*,[22] where the court explained the doctrine as follows:

> The cardinal change doctrine is a creature of the body of law which has arisen in the context of disputes over government contracts. A cardinal change occurs when the government effects an alteration in the work so drastic that it effectively requires the contractor to perform duties materially different from those originally bargained for. By definition, then, a cardinal change is so profound that it is not redressable under the contract, and thus renders the government in breach.

The cases above, with all their factual uniqueness, do permit some generalisation regarding cardinal changes. To summarise, two factors can identify whether a change is a cardinal change and then the contractor can base his total cost method on this requirement. They are as follows:

---

[20]   369 F.2d 701 (Ct.Cl. 1966).

[21]   210 Ct. Cl. 714 (1976).

[22]   773 F.Supp.335 (M.D.Fla. 1991).

1. *Quantitative Factors.* There are mainly three quantitative factors: (1) changes in size or amount; (2) changes in cost; and (3) changes in time. These factors are not equally significant. Changes in time are far less important than changes in size or cost. A sufficient change in size or cost, alone, may be sufficient. Most cases do not recognize cardinal change unless the size or cost overrun approaches 100%. There is also an interrelationship between the factors. If both size and cost increase substantially, but neither one to the 100% level, the combined effect of the increases still may constitute a cardinal change. In such a case, a significant increase in time may also tip the balance. If the cost and the size overruns do not exceed 50%, however, the odds of prevailing on a cardinal change defense are almost negligible. Finally, changes which reduce the amount of work or the expense of the contract generally do not constitute cardinal changes.
2. *Qualitative Factors.* By definition, these are more amorphous. Significantly, courts find qualitative cardinal changes even though the impact on job size or cost is moderate, or even minimal. Several factors, however, are discernable. The first is changes in fundamental structural design. This includes the deletion of buildings, significant changes to square footage, and changes in structural stress requirements. The second is work outside the scope of the contract. This includes research or design efforts not included in a basic manufacturing or construction contract. The third is fundamental changes in the construction methods. This includes abandoning specifications and requiring a trial and error approach, or changes requiring abandonment of onsite construction to fabricate at distant or expensive locations. Where qualitative changes such as these exist, courts find cardinal changes even when the project changes relatively little in size or in price.

Contractor, when faced with cardinal changes, can materialise their claim as a total global cost claim or when the cardinal changes are so profound they might try to attempt for the principles of *quantum meruit* or time at large. However, to claim under the global approach, the contractor is advised to approach the claim with the modified total cost approach where the claimant must eliminate from his quantum all costs that can be caused by his actions and all costs factors, that might hinder the success of his claim, such as overpriced bill of quantity, approved variations and change orders, and overhead costs that could be mitigated.

7.5 Damages: A Total Cost Approach

To be successful at his claims, it is necessary that the contractor damages must be adequately proved. The contractor must establish the amount of its damages with reasonable certainty and must also show that the damages were caused by the party from whom the relief is sought. The courts can be lenient when assessing damages if the contractor has used a formulae or a computational method to calculate these damages even if not accurately due to the impossibility to achieving the stated. Therefore, the use a total cost method or a total cost method can be applied by the claimant if proven that there is no other way available to prove the losses that have incurred.

In *Wood v Grand Valley R Co.*,[23] J. Davies had said as follows in applying the underlying principle of applying an approximate method in calculating damages:

> It was clearly impossible under the facts of that case to estimate with anything approaching mathematical accuracy the damages sustained by the plaintiffs, but it seems to me to be clearly laid down there by the learned judges that such an impossibility cannot relieve the wrongdoer of the necessity of paying damages for his breach of contract and that on the other hand the tribunal to estimate them whether jury or judge must under such circumstances do the best it can and its conclusion will not be set aside even if the amount of the verdict is a matter of guess work.

These standards of causation and reasonable certainty to calculate damages to recover sums alleged to have been lost in constructing a hospital project were explained by the court in *Wunderlich Contracting Co. v United States* (1965b) (*supra*) as stated by J. Cowen:

> A claimant need not prove his damages with absolute certainty or mathematical exactitude. It is sufficient if he furnishes the court with a reasonable basis for computation, even though the result is only approximate. Yet this leniency as to the actual mechanics of computation does not relieve the contractor of his essential burden of establishing the fundamental facts of liability, causation, and resultant injury.

J. Cowen continued:

> There is no exact formula for determining the point at which a single change or a series of changes must be considered to be beyond the scope of the contract and necessarily in breach of it. Each case must be analysed on its own facts and in light of its own circumstances, giving just consideration to the magnitude and quality of the changes ordered and their cumulative effect upon the project as a whole. The total cost plus profit theory of computing damages advanced here by plaintiffs is appropriate only in extreme cases, where no more satisfactory method is available.

In *Penvidic Contracting Co. v International Nickel Co. of Canada*,[24] a railroad contractor had undertaken to carry out track laying and surface ballasting to a 47.5 mile long railroad. The employer was in breach of its implied obligation to facilitate the work in a number of respects including failure to provide the necessary rail link to an existing railway for plant access. This factor alone resulted in revision to the whole method of construction whereby the contractor was required to commence work at a half way point and work in two directions. In the first instance, the contractor revalued the work claiming an additional increase in cost per ton on the contractual rate for top ballasting in compensation. His calculation was based upon the difference between the contractual rate per ton of ballast and the rate which he would have demanded had he foreseen the adverse conditions

---

[23] (1913) 16 DLR 361.

[24] (1975) 53 DLR (3d) 748 Can.

caused by the failure to provide the rail link. At trial, the contractor, due to having insufficient cost data, claimed damages for breach of the implied term assessed on the same basis as the revaluation of the work.

In *Penvidic v International Nickel*, the claim for damages was successful as J. Spence said:

> In an ordinary case, the plaintiff in an action for damages for such breaches of contract would prove the additional costs which it incurred. Under these circumstances, the plaintiff chose to put its claim for this extra ballasting on the basis of a claim for an additional sum per ton. That is the fashion in which it had attempted to have the respondent agree to pay extra compensation. That such an attempt ended in failure does not prevent the award of damages using the same measure as had been used in the vain attempt to obtain extra compensation." J. Spence then said that in support of the claimant method in calculating his damages: "I can see no objection whatsoever to the learned trial judge using the method suggested by the plaintiff of assessing the damages in the form of additional compensation per ton rather than attempting to reach it by ascertaining items of expense from records which, by the very nature of the contract, had to be fragmentary and probably mere estimations.

Under a global claim or total cost method, the contractor must calculate his damages with utmost certainty and must be fair and reasonable as the courts tempt to test this theory with precise reasoning. It is after all, the purpose of the claim is to put the claimant in a position he would have been in had those expectations been fulfilled. A claimant is not entitled to recover damages which would place him in a better position than it otherwise would have occupied had there been no wrong-doing by the defendant. This doctrine of betterment does not relieve the contractor of his essential burden of establishing the fundamental facts of liability, causation and resultant injury.

In a total cost approach the first step for the claimant is to base his baseline cost on the bill of quantity priced. The bill of quantity is the step stone and the basis for any calculation for a global claim under the total cost method or the modified total cost method. The bill of quantity must be, in certain circumstances, re-calculated and re-priced in order to eliminate from it discrepancies, errors, mistakes, omitted works and re-priced items. The bill of quantity should never be re-priced upwards as the global claim will re-compensate the claimant for all losses.

A clear tendering and bidding process prior to the commencement of a project will strengthen the claimant position as it shows that his original prices and quotes are competitive and the reason he was successful at procuring the job in question. Bidding is not just about solving the problem of choice of construction, but also makes clear the project price, schedule, quality and viability of the tender documents. Tenders in the engineering and the construction process are essential, effective as they prove the competiveness of the process. Tender bill of quantity is used to adapt to the reality the building of the entire project and makes the necessary preparations for a fair and transparent choice of a contractor to execute the job.

The second step is to calculate the total cost with precision by using all receipts, documents, purchase orders and full accounting and proper audit system.

The claimant must prove that he endeavoured his best to keep his costs competitive by showing his procurement methods and mitigation procedures. All costs that can hinder his method of proving losses and costs that are questionable with no viable records are best to be kept out and claimed separately or negotiated.

In a total cost method, the claimant must have established that the exact computation of damages is virtually impossible due to the complex interrelationship of the various factors attributable to delays and disruptions. However, contractors should prove that by standard, the best accounting records and job site documents are ordinarily used to prove the actual cost of any claim. One tool used by contractors in determining impact and losses is the critical path method. This tool allows the contractor to identify all activities required to perform the work. A delay in any of the activities along the critical path will cause a delay to the entire project. Therefore, these schedules can be helpful to a contractor in proving entitlement to additional time and the additional of time can be the basis to show the global loss incurred as a matter of multiplying the delayed time by the unit rate.

7.6   Baseline Cost: Bill of Quantity

The bill of quantity bidding method is commonly used in the international common practice, nearly a hundred years of history, with a wide range of adaptability and a more scientific, rational and practical method of choice of the contractor to construe the project. The bill of quantity role is for the bidders to bid for tenders, to provide a common basis of competitive bidding and is the basis for progress payments in the course of construction works. In addition, when engineering changes occur, the unit contract prices in the bill of quantity are an important reference standard for the claims.

As stated earlier, the bill of quantity represents the baseline the claimant bases his method of calculating the total cost incurred. The bill of quantity, also known as the project scale, is usually divided into units and divisions of general and technical specifications used in engineering calculation rules. The quantities and degree of accuracy depend primarily on the design depth with the corresponding drawings, but also relate with the contract form.

When adopted, the bill of quantity use in tendering is to fix pricing and makes basis for the calculation of the total losses and as a way to prove unit rates for variations or in the case of cardinal changes such as omitting parts of the work or increasing substantially the scope of work. The general practice of using fixed pricing is to define the amount of the projects, to set fixed direct costs and then in the form of rates for the calculation of indirect costs and finally to obtain the final offer.

Quantities in the bill of quantity draw large computing units and in the fine quality requirements are true reflections of the engineering practice in order to bring prices to the construction of the autonomy of trades and materials possible. In the project bidding process, unit rates, quantities and other documents such as drawings, specifications and technical documents used in the tender offer bid must take into account the content of the project itself, scope, technical requirements and features of the relevant provisions of the tender documents, project site conditions and other factors. It must also fully take into account many other factors, such as the contractor proposal to develop the total project schedule, construction programmes, subcontracting plans and resourcing plans.

The final quantities of work for an item may be different to the estimate in the tendered documents. The change in the final quantities of work for an item may so upset the balance of resources, plants, materials and the method of working, and eventually to make the unit price for the item inaccurate. The actual quantity of work for an item may differ from the estimate at tender for a number of reasons. In the case of excavation for instance the removal of unsuitable material or the extent of tunnelling in particular classifications of ground may only be estimated from ground investigation information and not known until work is carried out. Similarly the length of piles driven to a specified set may not be known precisely at each pile location. The change of quantities is dealt within the final cost analysis and is claimed separately or if not, it is included in the global claim.

The bill of quantity may serve a number of functions as:

- A breakdown of the tendered price, with no contractual status, but providing information for the selection from tenderers.
- An estimate measure of the work for the tendered price, to be used to arrive at a revised contract price once the actual quantities of work carried out are measured. This is the re-measure form of contract.
- A schedule of rates as the contract basis for valuing variations in the work.
- A basis for measure of the value of work completed for interim payments.

Contracts generally provide that the several documents forming the contract are to be taken as mutually explanatory of one another. There is no order of priority stated for the interpretation of the contract, so that the bill of quantity has the same status as drawings and specification. Some contracts, however, provide that the quantities set out in the bill of quantity are the estimated quantities of the work, but that they are not to be taken as the actual and correct quantities of the work to be carried out by the contractor. The claimant must verify that errors in description or omissions do not exist in the bill of quantity. Any such error or omission is corrected by the engineer reviewing and the value of work ascertained. Errors, omissions or wrong estimates in the description rates and prices inserted by the contractor are rectified before the bill of quantity modified cost is used as the baseline for the global claim.

The contractor must show that the bill of quantity is deemed to have been prepared and measurements made in accordance with the standard method of measurement referred to as '*Civil Engineering Standard Method of Measurement*' or other proven bodies. This provision is subject to an important proviso that general or detailed description or any other statement dos not clearly show the contrary. The method of measurement will specify the division of work into categories. In the building industry, the division is usually based on the basis of different trades and is generally very detailed. In the engineering industry, the division is usually less complex and composite items are used to describe the completed construction operation. There is normally a division for preliminary items such as mobilisation, site set up and insurances.

Standard methods of measurement have become increasingly more complicated. They give rise to claims for additional payment based on interpretation of the method. The tendency has been for the methods to provide detailed subdivision of work and therefore the scope for claims is based on ambiguities of interpretation, failure to measure the tendered bills in accordance with the method and the application of exceptions to measure. One, however, must be sure that any claims related to the method of measurement must be separated from the global claim or must be altogether disregarded in favour of the global claim.

The contractor must also verify that the actual quantities for an item do not differ from the quantity in the bill of quantity. If such difference so warrants, then the contractor determines the increase or the decrease of any rate tendered unreasonable or inapplicable inconsequence. As a matter of business efficacy, a term will be implied, in the absence of express terms, that the cost of the work for a bill item which has not been priced by the contractor is included in the prices entered elsewhere in the bill. As well, repetitive items must be omitted in the final calculation of the bill of quantity before using as the baseline for the total cost computation.

The contractor is not responsible for the increase or decrease in the quantity of any work, where it results from the quantities being different to those stated in the bill of quantity. Mistakes in the measure are corrected and are dealt with as compensation events and changes in quantities which are not minimal are also compensation events. Any mistakes in the bill of quantity which are departures from the method of measurement or are due to ambiguities or inconsistencies are corrected. Mistakes in the bill descriptions or quantities are unlikely to be remedied as a legal rectification of the terms of the contract to reflect the true intention of the parties. It is more likely than not, that the common intention will be that the tendered price should prevail, rather than a price revised to account of the error. Most standard forms of contract, which adopt bills of quantity, make provision to deal with errors in the bill descriptions and quantities, distinct from the effect of variations. All these points must be clearly stipulated to show the accuracy of the amount tendered and its usage when calculating the claim loss.

In summary, the bill of quantity must be generally re-priced and then applied as baseline in order to allow for the variations and the cardinal changes that have materially changed the quantities of the works. The change orders can be left out at this stage and to be included in the final cost analysis or to be claimed separately. The change in total cost of the project due to change orders is re-calculated based on the following:

1. Apply the contract rates unaltered to the changed quantities for the item of work.
2. Adjust the contract rates for the item of work, if the difference in quantities makes the balance of the rate inaccurate, leaving all other items including preliminaries unaltered.

3. Adjust the rates for other items of work, when planned execution is no longer valid due to the difference in quantity for the item of work.
4. Adjust the prices for preliminary items, which are affected by the difference in quantities. This will create difficulties unless the preliminary item is clearly time-related and the effect can be assessed on a time basis or if there is a build-up of the preliminary item prices.

When the project consists of different parts and zones and delays and disruptions vary between each part of the project, the bill of quantity must be restructured into smaller bills for each part. Then, the above list of valuations is done for each part. In certain instances, contractors try to provide a coefficient to each part based on the area of the part in relation to the total area and then calculate the losses based on the percentage used.

7.7 Total Costs Analysis

The amount to be claimed in a global claim is to calculate the final total cost and subtract from it the adjusted cost of the bill of quantity. The modified total cost method is a modification of the total cost method. In a modified total cost approach, the claimant must eliminate all damages that can be caused by his action or inaction. Therefore, the claimant is entitled to recover the reasonable value of the work performed by it less the payments made and less any costs incurred by the claimant itself which are not fairly attributable to the wrongful party.

The modified total cost method is the total cost method adjusted for any deficiencies and where the total cost method is used as only a starting point with such adjustments thereafter made to convince the court that the ultimate reduced figure fairly represents the increased costs the contractor directly suffered from the particular action of defendant which was the subject of the complaint. Other damages, caused by the wronged party, that can be computed and claimed separately must be eliminated from the global claim and claimed individually.

For contractors, implementing a system of final cost accounting appropriate to the size of a global claim is always a good idea. This is because the burden of proof to substantiate a claim always rests with the contractor. For judges, a proper accounting procedure should be required whenever permitted. Without it, the client may be held liable for claims based on contractor estimates that are inherently less reliable than actual costs.

Total cost accounting refers to the accounting procedures that a contractor uses to segregate its final costs to perform the work when subjected to delays and disruptions and a great number of change orders with no mean to segregate the cause and effect. Total and final cost accounting assists the parties determine the amount that the contract price should be adjusted for delays and disruptions, also known as a request for an equitable adjustment. As previously noted, the delays and disruptions require an adjustment to the contract price for an increase in the cost to perform the work. With proper accounting system, the task of determining the additional costs is largely one of extracting the data from the accounting system.

If the variations and change orders require new work that was not contemplated in the original contract, then the contractor should set up specific accounts to record labour, purchases of materials, intra-company transfers and other costs incurred to perform the new work. For example, suppose a contractor uses a job-costing system in which the client contract is coded 'A101', contractor working under that contract would charge their time to 'A101'. Personnel in the purchasing department would charge materials bought for that contract to 'A101' and so on.

A contractor calculating his final cost for a global claim under the total cost method or the modified total cost method without instituting actual accounting is at a severe disadvantage as the contractor may be unable to ascertain the actual costs it incurred to perform the changed work. Without actual costs, the contractor may be unable to prove their claim. Courts generally prefer that contractors prove their claims using the actual cost method as the actual cost method provides the court with documented underlying expenses, ensuring that the final amount of the equitable adjustment and not a windfall for either the client or the contractor.

Estimating actual costs may occasionally be used as an alternative. Such estimates may be based, for example, on contractor testimony as to the hours expended or on purchase orders for materials similar to those that were used. However, estimates are less credible than actual costs and are easily challenged.

To develop claims without proper accounting can be a difficult and costly process. Isolating specific costs after they have already been incurred can require, for example, a detailed review of payroll records, purchase orders of materials, modifications of subcontractor agreements, project progress charts and similar documents. Usually, a detailed cost analysis is too complex or time consuming to perform in-house. However, even if the contractor staff is capable of developing the necessary data, the effort required to do so and the resulting disruption to normal operations could exceed the cost that the contractor would have incurred to set up a relatively simple accounting procedure from the beginning.

The preferred way for a contractor to prove increased costs is to submit actual cost data because such data provides the court with documented underlying expenses, ensuring that the final amount is the equitable adjustment and not a windfall for either the client or the contractor. In order to calculate the final cost of a project the claimant must be fully aware of the importance to demonstrate his track record and of keeping precise supporting evidence and records that can support the claim.

Undoubtedly, admissible evidence in relation to an issue of valuation and assessment of damages should be particularised and costs highlighted. The sooner the cost information can be provided the better. Good accountancy practice on the part of the contractor can lead to early indication of difficulties on site, leading to earlier notification of loss. With an understanding of good construction accountancy practice, the following conditions are to be noted:

1. Identify what information is required.
2. Assess the reliability of the data.
3. Make the assessment and valuation in question.
4. There is an implied duty to co-operate and to act fairly and in good faith in relation to the valuation of the work.

Standard accountancy software is available which can facilitate the requirements of the industry. A basic working knowledge thereof, on the part of both the engineer in charge and the contractor, would be beneficial. One difficulty is that many firms, especially smaller contracting firms, maintain only very basic and disorganised accounts information. Accounts staff may, for example, be interested in record keeping for tax purposes only.

7.8 Evidence: Documents and Records

As demonstrated before, when a global claim is being made, it will only be the substantiated items on a list of grouped events that will be held as capable or incapable of supporting the financial claim being made. Similarly, it was shown that even though a claim was largely successful, the court will usually choose from a global claim, any part of it which is adequately substantiated, and use these parts to justify awarding that element of the claim. The elements of the claim for which nothing has been provided by the contractor by way of evidence however will be unsuccessful. While the court is willing to work with what it had to minimise the parts of the claim that would fall to be unsuccessful, not all courts will be as willing and the contractor should not assume it will have such an accommodating court, that will be willing to take measures to make up for shortcomings in the evidence supporting the claim.

These types of records that can be used by the contractor to support the claims being made were examined in the *Attorney General for the Falkland Islands v Gordon Forbes (Falklands) Construction Ltd*[25] case. The court had to examine the meaning of *'contemporary records'* as it was contained in the contract, which referred to the necessity of keeping these records to support a claim. The meaning given to the phrase was that it referred to original or primary documents, or copies of such documents and reference to generalisations, averages, percentages of productivity and similar figures were not acceptable. The court held that documents relating to actual figures of the actual case, prepared at the time of the disruption in question, and in response to it, and not later evidence brought forward for the purposes of the trial, such as witness statements, were what was necessary, particularly when the wording of the contract explicitly stated this.

---

[25] (2003) 19 Const LJ T1 49.

In regard the importance of backup evidence was highlighted by J. Sanders in *Attorney General for the Falkland Islands v Gordon Forbes (Falklands) Construction Ltd* as follows:

> The nature of the back-up evidence will obviously depend on the type of claim, but in almost every case detailed cost records and comparative programme/progress schedules will be necessary, together with references to correspondence, records of site meetings, site diaries and the like.

While it may be that the contractor will paraphrase and provide summaries of the original correspondence, meeting notes and other documents, and this can be done to highlight the point that the contractor is trying to make, it is vital that the actual documents are also available and that the other side to the dispute has the opportunity to examine and refer to the documents themselves and frame their own argument to defend their case. Paraphrasing such documentation accurately is of high importance as any party falsely paraphrasing such evidence is likely to be harshly penalised in the final view of the court.

Notwithstanding, the prudent contractor will be constantly vigilant for the types of situations described, and will give the earliest possible warning to the client, of his intent to claim and the anticipated grounds for doing so. In this way, under most contracts, the contractor is able to preserve his rights to claim until such time as the necessary information can be collected and appropriate analyses conducted.

Obviously, the extent of record keeping required for a particular construction job will depend on the type of contract. However, some record keeping will be required in any case because it is:

1. Required by law.
2. Required by the terms of the contract.
3. Needed to control the on-going work.
4. Needed as data for estimating future work.
5. Needed for preserving the contractor's rights under the contract.

The first item may be ascertained by referring to the authorities having jurisdiction over the place of the work. The second may be determined by a thorough reading of the contract documents, both in terms of the administrative requirements contained in the general and special conditions, and the technical requirements contained in the specifications. The third, fourth and fifth items are for the contractor to decide, and depend largely on his disposition.

A good set of records that might be kept on a fair sized construction project could well include the following files. Note that these files are assembled into blocks of like subject matter. This approach greatly facilitates ease of filing and subsequent recall. This list may seem like a lot of files and records, but most of them are kept by the well-organized contractor anyway:

1. Original contract tender documents and all subsequent revisions.
2. Instructions to contractor.
3. Contemplated change notices issued by the owner, change estimates and change orders received.

4. Subcontractor quotes, contracts, purchase orders and correspondence.
5. Shop drawings, revisions and re-submissions.
6. Shop drawing transmittals and transmittals log.
7. Daily time records.
8. Daily equipment use.
9. Daily production logs.
10. Material delivery records.
11. Accounting records: payroll, accounts payable and receivable, etc.
12. Progress payment billings under the contract.
13. Daily force account records, pricing and billings.
14. Contract milestone schedule or master schedule.
15. Short-term schedules and up-dates.
16. Task schedules and analyses.
17. Original tender estimate.
18. Construction control budget.
19. Actual cost reports.
20. Productivity reports and analyses.
21. Inter-office correspondence (all filed by topic).
22. Contract correspondence.
23. Minutes of contractual meetings.
24. Minutes of site coordination meetings.
25. Requests for information.
26. Notice of claims for delays and/or extra cost by contractor.
27. Government inspection reports.
28. Consultant inspection reports.
29. Accident reports.
30. Daily diary or journal entries.
31. Notes of telephone conversations.
32. Progress reports, weekly, monthly or quarterly.
33. Progress photographs.
34. Any other reports, such as special consultant reports.
35. A filing record of all the record files that are being maintained.

As well as managing the files, the records themselves also need managing. Some simple rules can help as follows:

1. Determine what records are to be kept, and how. Establish logs of the records, so that they can be found, referred to and/or followed up as required. Well-organised contractors establish standard reference lists and coding for all their contracts. This greatly facilitates managing, analysing and comparing contracts.
2. Once the records have been identified, ensure that they are in fact set up, maintained and used for managing the job.
3. Review the record-keeping system from time to time for better systemisation and structuring especially when they are large in volume. In addition, some records may become obsolete or redundant and should be discontinued. Unnecessary record keeping can waste a lot of time and money.

4. Records also take up space and equipment. Determine the useful life of the different components and take a systematic approach to record disposal.
5. Take steps to ensure accuracy, reliability and hence credibility. Unreliable records can be quite useless, as well as a waste of money, and possibly even detrimental.

7.9 Global Claims Avoidance and Mitigation

More formally, a contractor's claim may be defined as: '*A legitimate request for additional compensation (cost and/or time) on account of a change in the terms of the contract.*' It follows from this definition that a global claim may arise under any form of construction contract, except perhaps those very rare kind, in which all costs are fully reimbursable without any reservations at all. Of course, a claim is most likely to arise under a fixed price form of contract, and in fact today there are few such contracts in which there are no claims, negotiations and settlements before the contract is finally closed out. It also follows that it is essential to know exactly what is expected of the contractor under the terms of the contract both before signing the contract as well as during its execution.

This knowledge must not just be limited to senior management at the main office. Site supervisors who deal with the day-to-day work must be equally well informed. Strictly speaking, every article and requirement of the contract must be clearly understood, if the contents of the contract are to be faithfully carried out. It is a matter for great regret, therefore, that some contracts are written by lawyers in such a way that only other lawyers can understand them. Fortunately, the increasing use of standard documents and specifications has gone a long way to facilitate the expression of requirements, and thereby avoid disputes through simply misinterpretation. So three simple rules can be promulgated to avoid making claims:

1. Good knowledge of the contract provisions and requirements.
2. Proceed with the works diligently even in the case of disputes.
3. Proper documentation must be always implemented.

Typical sources of disputes claims are worth noting. Theoretically, any clause in the contract could become the basis of a claim. Indeed, it is a wonder that contracts have not become much simpler on this account alone. Generally, global claims may be identified as falling into one of the following main groups:

1. Changed conditions. Conditions different from that represented by the contract documents or known at the time of bidding on the work, such as different soil conditions, or unknown obstructions, etc.
2. Additional work. Disputes arise over the pricing and timing of additional work required, or even whether a piece of identified work is in the contract or not. Beware particularly of omissions in the design documents, requiring changes to

make a system work, especially if they appear in a subtle way through the shop drawing review and approval process.
3. Delays and disruptions. These refer to delays and disruptions strictly beyond the contractor control. They may be caused by the client directly, or by one of his agents. A prime example is failure to give access to the site of the work in a timely way or when an equipment is promised by the client and is not delivered on time. More frequently, delays and disruptions occur when working drawings are not provided in time to suit the work, or when hop drawings are not reviewed in a timely manner.
4. Contract time. Disputes often arise over a contractor's request for a time extension on account of changed conditions, required changes to the contract, or owner caused delays. Disputes may also arise over instructions to accelerate the work. Such instructions may not necessarily be explicit. For example, instructions to incorporate additional work without a corresponding time extension, especially if the work is on the critical path, is tantamount to an instruction to accelerate in order to meet the contract completion date.

Very often a contractor does not know the real cause for a claim until later after the events that have given rise to the situation. A typical case involves the accumulated impact of a series of changes, each of which may appear minor, but collectively have a disrupting effect out of all proportion to the work involved. Other changes may give rise to a re-scheduling of work, with consequent loss of productivity. Often, these impacts are difficult to determine till near completion.

Notwithstanding, the prudent contractor will be constantly vigilant for the types of situations described, and will give the earliest possible warning to the client of his intent to claim and the anticipated grounds for doing so. In this way, under most contracts, the contractor is able to preserve his rights to claim until such time as the necessary information can be collected and appropriate analyses conducted.

As noted earlier, for any contractor on all projects, records are required for estimating future work, and for protecting his contractual rights. Both of these require some form of post-contract review. However, there can be little argument that reliable data cannot be extracted from records created after the fact. Even the best of memories are fallible, and the written record serves to provide the solid reminder. Data may be extracted, analysed and presented in a different light, but satisfactory records cannot be created later.

Contractors can significantly improve their chances of recovering damages by keeping detailed and accurate records of their operations and of specific impacts of delay or disruption to their schedules. Several categories of records which are helpful in proving delay damages include the following:

1. Diaries.
2. Daily job reports.
3. Time records.
4. Accounting records.
5. Production records.
6. Photographs.

7. Charts.
8. Schedules.

Because changes and variations originate from the client and it is the contractor that suffers their effects, the contractor has to manage this risk so as to minimise his losses. The contractor should demonstrate that he carried out the following steps in order to make it possible to recover the costs of delays and disruptions:

- The contractor should research the contract documents thoroughly to confirm that a change condition actually exist, being a change order or a cardinal change.
- The contractor should prepare and submit a request for formal instructions giving the consultant a clear and detailed description of the changes in dispute.
- The contractor should provide notice of any likely effect on progress and on the completion of the project.
- The contractor should provide notice of any likely effects on preliminaries.
- The contractor should give notice of an intention to proceed to mediation, expert decision or arbitration[26] if the formal instruction on variations is not forthcoming.
- The contractor should put together a dispute file documenting the factual background and all responses or lack of them from consultant and owner.
- The contractor should inform the consultant of the need for any prerequisite determination by the consultant of the issue before adjudication or arbitration.
- The contractor should update its notices on the effects of time and cost as the variation work proceeds. Deadlines for issuing notices and variation orders must be monitored and implemented.

A schedule summarising all changes resulting from negotiations should be kept as well as related minutes of meetings between the relevant parties and signed by the parties present. Important things to note within the minutes of meetings are the instructions that were given to the contractor to enable him to make any price adjustments and the changes that have been agreed to by the parties. It is obvious from the above list that the contractor should proceed with the variation works despite the dispute.

An inclusion within the contract documents of all post bid documents and drawings, which were never given to the contractor in his bid process, can lead to drastic cost consequences later when decisions are required as to whether an item of work is a variation. Both parties should scrutinise and verify the individual documents forming the contract documents prior to award. Parties should take special care to include all the agreed changes to pricing, scope of works and division of responsibilities between them so that no ambiguity arises when construing the final total cost of the project. Documents such as the design and construction programs, progress reports, drawing release dates and approval schedules must be constantly updated so that reliance on the information contained therein does not become obsolete in the event of litigation later down the road.

---

[26] Depending on the dispute resolution provision in contract.

The contractor has a duty to show that he has done all necessary steps to mitigate the effect of losses. In *British Westinghouse v Underground Railway Co.*,[27] Lord Haldane, by establishing this rule, held that the wronged party has a duty "of taking all reasonable steps to mitigate the loss consequent on the breach, and debars him from claiming any part of the damage which is due to his neglect to take such steps."

There have been various debates on how far the common law duty to mitigate is to be imposed but eventually the courts seem to have settled on the following principles:

- It is not a duty to mitigate loss but the extent of liability on the part of the other party that is reduced because a defendant can only be liable for such part of the plaintiff loss that has been probably caused.
- The extent of the mitigation required is a question of fact and not law.
- The limitation to the extent of mitigation is reasonable steps.
- There is no need to embark on an uncertain or risky step in mitigation or one that may cause a loss of reputation.
- The onus is on the defendant to proof the failure to mitigate.
- Any cost incurred on embarking on a reasonable mitigation process will also be recoverable against the defendant.

However, any gain resulting from the plaintiff reasonable steps in mitigation must be balanced against the loss caused by the breach and any loss resulting from such reasonable steps is recoverable.

References

Case Law

Allied Materials and Equipment Co. v United States 210 Ct. Cl. 714 (1976)
Amelco Electric v City of Thousand Oak (2000), 82 Cal.App.4th 373
Atlantic Dry Dock Corp. v U.S. 773 F.Supp.335 (M.D.Fla. 1991)
Attorney General for the Falkland Islands v Gordon Forbes (Falklands) Construction Ltd (2003) 19 Const LJ T1 49
Biemann and Rowell Co. v Donohoe Companies Inc 556 S.E.2d 1, 5 (N.C. Ct. App. 2001)
Boyajian v United States 423 F.2d. 1231 (U.S. Ct. Claims, 1970)
British Westinghouse v Underground Railway Co. (1912) A.C. 673 at 689
Ernst EC Inc. v Koppers Company, Inc. 47 F. Supp. 729 (WD Pa. 1979)
Dillingham-Ray Wilson v City of Los Angeles (2010) 182 Cal.App.4th 1396
General Contracting and Construction Co. v U.S. 84 Ct.Cl. 570 (1937)
Shupe GM Inc. v United States, 5 Cl. Ct. 662 (1984)
Huber, Hunt and Nichols Inc. v Moore (1977) 67 Cal.App.3d 278
Crosby J and Sons Ltd. v Portland Urban District Council (1967) 5 BLR 121

---

[27] (1912) A.C. 673 at 689.

Hedin JD Constr. Co. v United States, 347 F.2d 235 (Ct. Cl. 1965a)
Lichter v Mellon Stuart Company 305 F.2d. 216 (3d Cir. 1962)
London Borough of Merton v Stanley Hugh Leach (1985) 32BLR 51
Luria Brothers and Company, Inc. v U.S. 369 F.2d 701 (Ct.Cl. 1966)
McDevitt and Street Co. v Department of General Services (Fla. 1st DCA 1979)
Saddler PL v U.S. 287 F.2d 411 (Ct.Cl. 1961)
Penvidic Contracting Co v International Nickel Co of Canada (1975) 53 DLR (3d) 748 Can
Phillips Construction Co Inc v United States 394 F 2d 834 (1968)
Propellex Corpration v Brownlee, 342 F.3d 1335 (2001)
Servidone Construction Corporation v the United States 931 F.2d 860. April 24, 1991
Wood v Grand Valley R Co (1913) 16 DLR 361
WRB Corp. v United States 183 Ct. Cl. 409, 426 (1968)
Wunderlich Contracting Co. v United States, 351 F So. 956 (Ct. Cl. 1965b)

Books

Callahan M (2010) Construction change order claims. Aspen Publishers, Maryland
Carnell N (2005) Causation and delay in construction disputes. Blackwell Publishing, Oxford
Davison P, Mullen J (2008) Evaluating contract claims. Wiley-Blackwell Publishing, Oxford
Pickavance K (2010) Delay and disruption in construction contracts. Sweet and Maxwell,
    London, UK
Reese C (2010) Hudson's building and engineering contracts. Sweet and Maxwell, London

Chapter 8
Global Claim: A Case Study

8.1 Introduction

This is a global claim relating to a project (called for the purpose of this claim Project A) which is located in the Gulf region. It is fictitious and all description, data and facts stated are made up by the author and only intended to provide the reader with an insight how to write a global claim. This claim is classified as a modified total cost claim, where the costing methodology is different from a typical global claim and where the claimant only claims for the costs that cannot be quantified and the amounts claimed relate to delays and disruptions caused by the client solely.

The approach used in this chapter is to provide the readers with the know-how in preparing, writing and calculating global claims. The author has used all his knowledge, past experience and handling numerous claims in order to combine and provide as much generic and informative information as possible to prepare this chapter. It is not meant to be that the claim presented is what claimants should copy and use exactly in the same format when preparing their claims.

Each claim is different in the way it is formulated, structured and how the data is accumulated and even though the calculation methodology sounds simplistic for global claims, this is not true. Each claim should be prepared from scratch while the fact, data and relevant documents impose themselves on the team preparing the claim so they navigate their way into the uncharted waters of the findings. A good claim leader is the one who will lead them into producing the final product. The task is formidable with claims that can amount into hundreds and millions of US Dollars and the documents, data and drawings run into tens of thousands and even more.

The claim leader and his team have to consider the time frame that they have to work within and of course the expenses in preparing a claim. Therefore, a solid and efficient mechanism of data, a systematic review of all documents and the best approach used to prepare the claim are essential.

The other point that is essential in preparing global claims is who should be preparing the claim. A claim, where most of the documents are technical and the

A. D. Haidar, *Global Claims in Construction*,
DOI: 10.1007/978-0-85729-730-3_8, © Springer-Verlag London Limited 2011

claim falls within the ambit of the contract, requires a team consisting of quantity surveyors and engineers which can be sufficient. If the claim is relying on doctrines in law falling outside the scope of the contract, lawyers, arbitrators and engineers aware of the applicable laws, as well as quantity surveyors, are essential members of the team preparing the claim.

8.2   Description of the Claim

This claim includes the legal and contractual background that the Contractor is basing its right for compensation and which has imposed a heavy burden on the operation of the Project and placed the Contractor under extreme economic duress due to the heavy losses that he has incurred. Project A was marred with numerous problems from the beginning when the Contractor mobilised his staff, workforce and resources. The number and impact of these problems are rarely encountered in one single project and if only one or two of these problems occur in one particular project they can inflict heavy losses on the Contractor. Each claim heading is written in detail and all the legal background, the problems encountered on the project and the amounts to be claimed are supported with facts and documents including letters, reports, invoices, tender documents, subcontractors correspondences and variation orders.[1] The Contractor is seeking remedies to the problems encountered in the Project and relief though common and applicable laws by applying certain doctrines such as misrepresentation, economic duress and frustration which allow a degree of release of the terms of the contract or adjustment of these terms in the face of the exceptional unforeseen conditions that occurred through the life of the project.

The delays and obstacles that occurred could not be foreseeable by the Contractor or any other experienced contractor and where outside the scope of the assumed risks considered under the contract at the time the job was awarded on the 3 November 2006 and during the period prior to that date when the job was estimated and the plans to construct were put in place. The consequence of the above disruptions is that the contract could be rescinded due to the doctrine of frustration where, in the event of frustration, the contract will be discharged in relation to future performance obligations of both parties. The Contractor even knowing its right to rescind the contract due to these circumstances decided to continue the work even though under heavy losses amounting to more than 42 million USD and where the project could have been delayed beyond any reasonable date that could have been envisaged and foreseen if not for the Contractor acceleration *operatus*.

---

[1] These facts and documents are usually of a sheer volume and could be not included in this book.

The Contractor faced enormous difficulties from the outset and these are explained in great detail in the claim document. A summary of these difficulties falls under the following headings:

- Late arrival of the consultant's team to site.
- Effect of the construction boom in the Gulf region.
- Effect of the remoteness of the site.
- Effect of dealing with the client structure, the site safety and the security regime.
- Disruption arising from change orders and disputed change orders.
- Effect of delayed approvals of the drawings.

The changes amount to a fundamental change in the scope of the work and that the additional costs were unforeseeable. The Contractor was denied the costs of the cumulative impact which he can rightfully claim and as he is clearly entitled to additional compensation as a result of the numerous change orders that had cumulative impacts throughout the execution of the project. Notwithstanding the lack of absolute direct causal proof, it is clear—on the balance of probabilities—that for the '*but-for*' test:

- The client numerous, pervasive and cumulative changes instructed throughout the execution of the works.
- The client onerous security procedures and bureaucratic requirements.
- The remoteness and harshness (social and environmental factors) of the site.
- The consultant inefficiency and lack of necessary staff.

The Contractor would have been able to complete the works within budget but for the client caused delays and disruption and as such, as detailed in this global claim, the Contractor is entitled to compensation for cumulative disruption 42,684,293 USD.

8.3 Contract Particulars

The form of contract used is a bespoke contract provided by client. The following is a list of the relevant contract clauses, which have relevance to this claim document:

Schedule 'A' General Terms and Conditions

Clause	Titles
1.	Definitions
2.	Contractor obligations
7.	Company obligations
8.	Work schedule and progress reports
9.	Work commencement, execution and completion
10.	Changes
19.	Claims settlement; disputes

Schedule 'A' Contract Price and Payment Provisions

Clause   Titles
1.   Contract price
2.   Compensation for change orders

The following are the pertinent aspects of the Contract:

Client	:	Gulf Operations Company Limited
Contractor	:	Prime Contractors Limited
Procurement Method	:	Select tendering
Contract Signing Date	:	3 November 2006
Type of Contract	:	Lump sum
Contract Price	:	USD 61,987,185
Order to Commence given	:	28 November 2006
Completion Date	:	28 October 2010
Performance Bond	:	5% of the Contract Price
Advance Payment	:	5% of the Contract Price
Period of between Interim	:	One month, submitted within ten days of month with payment within the first 15 days of the following month.
Retention	:	5%
Liquidated Damages	:	First 14 days—0.05% of the Contract Price
		Second 14 days—0.10% of the Contract Price;
		Thereafter—0.15% of the Contract Price
		Up to a maximum of 10% of the Contract Price
Defects Liability Period	:	One year
Language	:	English

8.4 Summary of Events

The parties of the contract to which this claim is subjected to are as follows[2]:

1. Gulf Operations Company Limited, called 'Client' or 'Company'; and
2. Prime Contractors Limited, called 'Contractor'.

Project A is an extensive construction project completed in a remote location in the Middle East Gulf region. The Project value is 61,987,186 USD and is being constructed in three phases over a construction period of 1,396 days. This claim is

---

[2] This is a fictitious case study. All names, details and data have been altered. The global claim described in this book is based on other claims with all details changed to hide and protect all sensitive data. Copyright, where necessary, has been obtained.

prepared after the completion date of the project and the Contractor has submitted all necessary notices in due time to preserve his right to submit this claim.

This is a global claim relating to Project A with a value of 42,684,293 USD. The Contractor entered into contract with the Client on the 3rd of November 2006, to construct the overall development. The contract commenced on the 28th November 2006 and was due for completion 1,396 days later on the 28 October 2010 for a value of USD 61,987,185.

The Contractor faced enormous difficulties from the outset and these are explained in great detail in this claim document. However, a synopsis of these difficulties can be examined under the following headings:

1. Late arrival of the consultant's team to site.
2. Effect of the construction boom.
3. Effect of the remoteness of the site.
4. Effect of dealing with the client structure, the site safety and the security regime.
5. Disruption arising from change orders and disputed change orders.

Notwithstanding the mentioned reasons and causes of delays and disruptions caused by the Client, it is clear that the Contractor would have been able to complete the works timeously, without added resources, and within budget but for the Client caused delays and disruption, as such, the Contractor is claiming entitlement to compensation for cumulative disruption of 42,684,293 USD.

Late Arrival of the Consultants Team to Site—It is imperative that projects of the size and complexity of Project A to gain early momentum in order to achieve a rate of output that will result in the execution and the completion of the works in accordance with the planned schedule envisaged during the tender process. Essential to this aim is the early and proactive involvement of the consultants. Unfortunately, for Project A, the arrival of the consultants site-based teams, in the numbers required to facilitate the necessary early momentum, was very slow and this had a dramatic negative effect on the Contractor's planned rate of output in the early months. A proactive consultant approval of the Contractor drawing and material submittals is essential to facilitate the Contractor need to gain this early momentum. Unfortunately, as it is clear from the body of this claim document, the consultants failed to provide the early assistance required. Apart from the drawings and materials approvals, which are essential in the early phase of any project of this size and complexity, the consultant team is needed in sufficient numbers to address the many queries the Contractor will formulate in the early phases of the works. In addition to addressing the many queries, the consultants will be required to issue many drawing clarifications to ensure that the Contractor can commence early construction activities in a meaningful and purposeful manner.

Effect of the Construction Boom—The Gulf region experienced an unprecedented construction boom during the period of 2006–2008. Labour and materials costs escalated at exponential rates which were completely unforeseeable at the time of tender. Mega projects costing billions of dollars became almost common place. The demand for resources became insatiable and the supply of resources, which are relatively inelastic, became increasing difficult to secure. With demand outstripping supply in this way, labour resources became very mobile and relocated regularly tempted by the ever appreciating terms offered by contractors to secure their resource needs.

This phenomenon had two dramatically negative effects on Project A. Firstly, the tender allowances for resources were fully consumed far earlier than planned and secondly, the difficulty in securing sufficient resources to satisfy the requirements of the planned activities severely degraded the Contractor planned output and progress of work. This claim will not address the prohibitive increase of materials and labour costs as this will be claimed separately as it can be quantified. The disruption of obtaining the required resources is however taken into consideration.

Effect of the Remoteness of the Site—The factor of the remoteness of the site is closely linked to the construction boom because fierce demand for labour resources meant that the labour pool could be more selective in terms of choosing more palatable locations. Project A location is in an inhospitable location in the middle of the desert and a long-distance drive from the main centres of population in the Gulf region and therefore the labour pool was required to live in prefabricated camps adjacent to the site with limited recreation facilities. As the construction boom roared in the region it became increasingly more difficult to attract the required labour resources to this location.

The direct and indirect costs became more significant as a result of the knock-on effect on the costs of out of sequence work and considerable rework caused by the disrupted labour availability and the ever increasing cost of obtaining willing labour to work in this hostile environment. There was an unprecedented movement of labour in and out of Project A which severely disrupted the required level of output. Consequently, the difficulty in securing sufficient resources to satisfy the requirements of the planned activities severely degraded the Contractor's planned output and progress of work. These factors were so extreme that they border on the doctrine of *force majeure* as the construction expansion that the Gulf has seen during this period was unforeseen and unexpected for any professional working in the construction industry. The Contractor had to increase his workforce by more than 50% than his planned schedule to allow for this high movement of labour. It was a market where labour was dictating the planned activity of this Project.

Effect of Dealing with the Client Structure, the Site Safety and Security Regime—The Contractor had no comprehension of how bureaucratic the Client review structure and the site safety and the security structure would be. The contract documents were, at best, ambiguous about this fact. In reality, there is a measure of misrepresentation as to the *modus operandi* of the site safety and security regime procedures. Layers of bureaucratic red tape had to be navigated to attain simple tasks relating to receiving instructions, entering the site, obtaining passes for labour and subcontractors, and to abide generally with the site safety and the site security procedures imposed by the Client. What was considered routine procedure in estimating the effort to perform in accordance with the contract was in reality a nightmare that depressed staff morale, produced expensive out of sequence work, reduced productivity and caused rework to an unprecedented level. In brief, the effect of this imposed regime of bureaucracy was to disrupt and delay planned output and construction activities. This level of bureaucracy should have been made apparent in the tender documents to allow the Contractor sufficient data to plan and price accordingly. No amount of explanation would have provided an adequate description of the difficult conditions that the Contractor faced on this project due to the security and working conditions imposed by the Client.

In addition to the direct difficulties associated with dealing with the Client structure, the disruptive effect of the site safety and security regime procedures on site moral, productivity and effectiveness were significant. The requirement for the Contractor to conform and to accept Company imposed changes was invariably dictated by applying coercive pressures through these demanding procedures.

Another aspect of the structure, the site safety and the security regime involves elements of frustration. The implicit intention of the site safety and the security regime was, for the Contractor, a Client requirement to ensure safety and security not the basis to frustrate the Contractor in implementing what was clearly a very onerous set of construction tasks. The loss caused by the unexpected contingency of dealing with a very demanding and inflexible bureaucracy was very significant indeed. Preliminary estimates of losses caused by these stringent measures using dynamic simulation have indicated an order of magnitude ranging from 80 to 110% of the as-built costs incurred by the Contractor. When the obligation for the Company to provide site safety and security regime became one in which it was impossible or completely different from what the Contractor had originally planned, then the doctrine permits the frustration of contracts: "Where, after a contract is made, a party's performance is made impracticable without his fault by the occurrence of an event the non-occurrence of which was a basic assumption on which the contract was made, his duty to render that performance is discharged, unless the language or the circumstances indicate to the contrary."

Although, the legal rule governing the doctrine of frustration states that in the event of unplanned contingencies—which in this project are the site safety and the security regime which effectively became worse as the project evolved—there was enough justification for rescinding the Contract. The Contractor has decided, in good faith, to continue committing its goodwill and resources in completing this prestigious though very difficult project against all odds.

Disruption Arising from Change Orders and Disputed Change Orders—A considerable volume of change orders were issued during the construction period and these changes, combined with the very considerable number of disputed change orders, as detailed in the body of this claim document, caused considerable cumulative disruption to the Contractor's regular progress of work. The Contractor asserts that the number, timing and effect of the changes issued impacted his ability to plan and perform the work. The existence of unplanned events and conditions is proved beyond any reasonable doubt. The Client, by issuing this great number of change instructions disrupted the works intensely. The cumulative impact of the many Client initiated changes over the duration of the works, which was beyond the Contractor's control and foreseeability is clearly the most probable cause of the delay and the ripple disruptive effects which have impacted the disruption and great losses the Contractor has suffered during the execution of Project A.

Multiple change orders have been issued on this project; and there is a solid *prima facie* case showing that the Client's numerous change orders form a basis for the Contractor to globally claim as it is impossible to differentiate between their cost impacts or to calculate some of the change orders individually due to their complex interdependency. Where it was possible to do so, the Contractor has highlighted the cost of impact of the separate variations to be deducted from the total loss.[3] The Contractor asserts that the execution of Project A could have been so much lengthier, if not for the acceleration exerted by the Contractor by adding resources, more disrupted and costlier than the Project was originally agreed upon in the contract and anticipated at the time the bid was submitted.

The Contractor asserts that:

- The Client disruptions caused the material change in the nature, scope and schedule of the Project and the Contractor suffered the effects of the events, the cumulative impact in the productivity and the heavy losses due to the mentioned disruptions.
- The changes amount to a fundamental change in the scope of the work and that the additional costs were unforeseeable.

The Contractor was denied the costs of the cumulative impact which it can rightfully claim. As such, the Contractor is clearly entitled to additional compensation as a result of numerous change orders and the other mentioned reasons that had cumulative impacts throughout the execution of the Project.

8.5 Claim Components

In the disruption associated with the cumulative changes and causes for the construction of Project A, it is clear that:

1. The cumulative impact of the excessive changes and the disruptive causes in the Project affected the performance and schedule of works significantly.
2. The cumulative impact of the excessive changes and disruptive causes increased the cost of performance.
3. The impact was not foreseeable when the disruptions occurred and the change orders were acknowledged.
4. The changes were so numerous and the disruptive causes were overlapping that it constituted reasoning for this global claim with a modified approach.

---

[3] This process is called modified total cost method.

5. The agreed changes were so numerous or overlapping that the Contractor obtained less than a full recovery for the individual changes when approved.
6. At the time that the Contractor negotiated the changes approved by the Client, he had no way of knowing the full impact of the changes and the disruptions on the Project overall final cost.

Cumulative Effects of Unforeseen Events and Conditions—The extension of time component of the cumulative effects of the disruption events had a great significance in that the Contractor was delayed beyond any reasonable date to complete Project A. This extension of time was mitigated by the Contractor through his acceleration procedures. The buildings in Phase 1 generally commenced in accordance with the planned commencement date; but most of these buildings suffered severe disruptions along with the inevitable delays to the completion dates.

The buildings in Phase 2 generally did not commence in accordance with the planned commencement date and, in addition, most of these buildings suffered severe disruption with the inevitable delays to the completion dates. The Contractor was unable to make these sites available in accordance with the express requirements of the Contract as his recourses were unavailable due to accelerating production in Phase 1.

The buildings in Phase 3 are a mixture of buildings that some did not commence in accordance with the planned commencement date and others which started approximately at the planned date. In addition, most of these buildings suffered severe disruptions with the inevitable delays to the completion dates. The Contractor was unable to make these sites available in accordance with the express requirements of the contract as his recourses were unavailable due to accelerating production in Phase 2.

Effect of Delayed Approvals for the Drawings and Materials Submittals—The successful execution of any construction project depends largely on the efficiency of the approval process of the shop drawings and the materials submittals. On any project such as Project A, with drawing and material submittals running into many thousands, it is imperative that the consultants respond to submittals in a reasonable timeframe. Failure to respond in a reasonable timeframe will delay and disrupt the Contractor planned progress of work and will be detrimental to the Contractor objective of completing the works within the time for completion.

Unfortunately, the actual history of the consultants response time on Project A was very poor and was an important contributory factor to the disruption and delay experienced by the Contractor. The Contractor makes this assessment on the basis of analysing the return of submittals within three categories which are (1) Greater than 15 days but less than 25 days; (2) Greater than 25 days but less than 50 days; and (3) Greater than 50 days.

With respect to the first category 7,842 drawing submittals took more than 15 days for the Contractor to receive a response; with respect to the second category 3,586 drawing submittals took more than 25 days for the Contractor to receive a response; and with respect to the third category 1,824 drawing submittals took more than 50 days for the Contractor to receive a response.

Similarly, with material submittals, there were excessive delays in receiving a response from the consultants appointed by the Client. With respect to the first category 835 material submittals took more than 15 days for the Contractor to receive a response; with respect to the second category 482 material submittals took more than 25 days for the Contractor to receive a response; and with respect to the third category 152 material submittals took more than 50 days for the Contractor to receive a response.

The enormous difficulties detailed above on this project have resulted in the Contractor incurring extremely heavy losses. The Contractor proffers that the Client is duty bound to compensate the Contractor for the extremely heavy losses incurred and effort it took for the drawings approval and materials submittals approval on this project for reasons which were completely outside the Contractor control and completely outside the Contractor contemplation at the time of submitting the tender.

Furthermore, as it is clearly documented in the rest of this claim document, the Contractor contends that he had the right to seek relief under applicable laws from the unbearable losses described above. The elements of the law which are pertinent to this particular contract are:

1. Misrepresentation;
2. *Force Majeure*;
3. Frustration; and
4. Economic Duress—including element of Coercion.

The Contractor contends that the contract could have been rescinded under at least one, if not all three, of these legal concepts. Rescinding the contract for reasons of misrepresentation, *force majeure*, frustration or economic duress would have allowed the Contractor to cut its losses in the early stages of the contract and to have realised an overall loss of a fraction of the losses it incurred at the end of the project. Rescinding the contract early would have minimised the Contractor exposure to loss but it would have in turn exposed the Client to considerable additional costs. The Project would have been re-tendered, or negotiated with the next lowest Contractor, and the Client would have faced considerable delays and additional costs and losses.

However, the Contractor valued its long-standing relationship with the Client, and its impeccable record in the Gulf region, and as a result it did not pursue the way in rescinding the contract. It did so trusting that an amicable settlement would be reached with the Client when the works were nearing completion.

The total losses the Contractor is hereby submitting for compensation are listed in the following table. The Contractor has adjusted his losses by subtracting from this claim the variations agreed to be paid by the Client which amount to 394,870 USD and the other claim that he will be submitting in relation to the materials cost increase which will total to 2,440,480 USD. Therefore this global claim amount falls under the modified total cost method and is summarised as follows:

Total losses = 45,519,643 USD.
Agreed variations = 394,870 USD.
Material escalation claim = 2,440,480 USD (This claim will be submitted separately).
Global claim amount = 42,684,293 USD (This is the amount submitted in this claim; Table 8.1).

8.6 Legal Analysis

Project A was marred with delays from its starting date and was subjected to many unforeseeable factors including design errors and omissions, a large number of client initiated changes and variations, unanticipated obstacles due to site restrictions, restrictive bureaucratic practices and unforeseen lack of resources that are acts bordering on *force majeure*. Factors such as the unexpected lack of labour and increase in costs together with strict site restrictions and access have imposed a heavy burden on the operation of the Project and have put the Contractor under extreme economic duress due to the heavy losses that he has incurred.

The Contractor is relying on its awareness of the general policy of the law towards commercial transactions including construction contracts which is based on freedom of contract within an established framework of rules. But the Contractor would like to emphasise that this is balanced by countervailing policies, some of which might be explained on the grounds of either public policy or doing justice. The *Unfair Contract Terms Act 1977* in the UK, similar Acts in European countries and the *contra proferentem* doctrine are all concerned to do justice by countering unequal bargaining power and promoting fairness in contracts, and that the consideration payable amount in the absence of express agreement shall be the reasonable payment in the amount of USD 42,684,293.

This is quite evident in this global claim where the Contractor is making demands for loss and expense as reasonable payment to compensate for damages that accrued in the Project where the contract terms unreasonably prevented the Contractor from doing so. The courts have intervened by imposing liability in the absence of, or alongside, contractually binding obligations and the Contractor hereby is relying on the stated legal doctrines for relief against the binding contractual obligations that has forced him to work under extreme duress.

In this claim, the Contractor stresses that there are other areas where the law intervenes to provide relief or remedies despite apparently binding agreement or in the absence of binding agreement such is the case in this particular contract drawn between the Client and the Contractor. Where the courts intervene to upset apparently binding agreements, they do so to preserve the rights of the parties that had endured economic duress such in the case of the

Table 8.1 Breakdown of
global cost claim by structure
component

Type	Claim Amount USD
Main Area—Phase 1	
Project A Building B100	19,109,835
Site Development	2,049,146
Chiller Plant	3,180,371
IT Building Coastal Area—Phase 2	
Security Offices 1	654,762
Restaurant	158,412
Officer Quarters	114,301
Mess & Recreation Building	237,561
Security Offices 2	42,474
Security Offices 3	355,100
Gulf House	106,370
Site Development	292,000
Generator Room	287,083
IT Building	2,984,372
Site Development	313,011
Office Building—Phase 3	
Generator Room	257,117
Main Building-B159	1,426,129
Site Development	694,495
Main Gate House	297,239
Demolishing	310,911
Govt Office	70,711
Office Building-B120	825,148
Drilling & Maintenance Shop	54,701
Covered Parking	2,807
Well Testing Shop	617,875
Shaded Yard 1	12,835
Drilling Chemicals & Mud Store	1,678,803
Central Laboratory Building	3,347,190
Central Chemical Store	35,411
Generator Room	205,083
Pump Room	798,089
Site Development (Area 1)	1,648,831
Material Test Laboratory	240,747
Site Development (Material Lab.)	121,531
Building Extension	2,634,532
User Store 1	111,945
User Store 2	116,505
Shaded Yard 2	126,210
TOTAL	**45,519,643 USD**

Contractor in this particular project. Freedom of contract as been obliged by
the parties to this contract is the paramount contractual policy, and most of the
disruptions and duress that the Contractor was subjected to are deviations on
that principle.

The point made by J. Jessel in *Printing and Numeric Registering v Sampson*[4] at p. 465, to a plea that a contract should be held unenforceable on the grounds of economic duress, but they are equally valid in other areas which involve the courts in declaring a contract void, voidable, unenforceable or discharged before performance is completed, is clearly stated when he said:

> It must not be forgotten that you are not to extend arbitrarily those rules which say that a given contract is void, because if there is one thing which more than another policy requires, it is that men of full age and competent understanding shall have the utmost liberty of contracting, and that their contracts when entered into freely and voluntarily shall be held sacred and shall be enforced by courts of justice. Therefore you have this paramount policy to consider—that you are not lightly to interfere with freedom of contract.

The doctrine was recognised by the Privy Council in *Pao On v Lau Yiu Long*[5] where the possibility of economic duress as grounds for setting aside the contract was acknowledged. It was also recognised and applied by the House of Lords in *Dimskal Shipping v ITF, The 'Evia Luck'*[6]:

> A contract entered into under duress is voidable and void. A person who has entered into the contract may either affirm or avoid such contract after the duress has ceased; and if he has so voluntarily acted under it with the full knowledge of all the circumstances he may be held bound on the ground of ratification, or if, after escaping from the duress, he takes no steps to set aside the transaction he may be found to have affirmed.

Therefore, by understanding that this contract has been entered into by the parties freely, the Contractor has done his utmost not to declare this contract void, voidable or unenforceable even though he has been working under tremendous duress and incurring losses that other contractor could not sustain. This fact and others could render any type of contract under the ambit of law and courts as dischargeable.

8.7 Applicable Laws Related to this Claim

The above factors that hindered the progress of the works and caused tremendous damages to the Contractor during the course of the project, under law, allowed a degree of release of the terms of the contract or adjustment of these terms in the face of the exceptional unforeseen conditions that occurred through the life of the project till present. The Contractor failure to comply with the terms of the contract, or even accepting the penalties imposed in rescinding the contract, could have its penalties and liquidated damages to comprise an amount to no more than USD

---

[4] (1875) LR 19 EQ 462.

[5] (1980) AC 614.

[6] Dimskal Shipping Co SA v International Transport Workers Federation, The Evia Luck (1991) 4 All ER 871.

6.2 millions.[7] The Contractor was working under huge, ongoing losses that amounted to 42,684,293 USD, which is beyond any reasonable amount that should or could be incurred under the contract; and reasons for delays which have effectively made the Project time at large.

This case, therefore, raises sharply the question to what is the nature and extent of the duty of the Client whose contract operations has caused heavy losses to the Contractor executing the works. The Contractor in this instance raises the notion that the Client must not carry out or permit the continuation of an operation which he knows or ought to know clearly can cause such losses, however improbable that result may be, and that the Client is only bound to take into account the possibility of such damage, if damage is such that a reasonable Client careful of the operations of the project being executed, that he would regard as material.

The test to be applied here is whether the risk of damages to a contractor was small that a reasonable client in the position of the existing client, considering the matter from the point of view of profitability, would have thought it right to refrain from taking steps to prevent the losses. In considering the matter, the Client's right to take into account, not only how remote is the chance that the Contractor might lose a great amount of money to make his whole business in jeopardy, but also how serious the consequences are likely to be for the Contractor losses, to take into account the difficulty of the remedial measures.

The delays, disruptions and obstacles that occurred could not be foreseeable by the Contractor or any other experienced contractor and where outside the scope of the assumed risks considered under the contract at the time the job was awarded in October 2006 and during the period prior to that date when the job was estimated and the plans to construct were put in place. The consequence of the above disruptions is that the contract could be rescinded due to the doctrine of frustration where in the event of frustrating the contract can be discharged in relation to future performance obligations of both parties.

The Contractor, even knowing his right to rescind the contract due to these circumstances decided to continue the work even though under heavy losses amounting to more than 40 million USD and the project being could have been delayed beyond any reasonable date that could have been envisaged and foreseeable if for the acceleration procedures carried out by the Contractor to complete within a reasonable time considering the delays and disruptions caused by the Client.

Insofar the courts have been willing to provide relief against disadvantageous contracts and they have done so by coordinated steps with common philosophy and principle. Where law has intervened, the approach has generally been universal in providing relief and remedies to the injured party. Therefore, the Contractor is insisting that the contract, they have entered into in good faith, is disadvantageous and the terms of the contract are not fair even though negotiated

---

[7] This is based on the liquidated damages clause in the contract which amounts to no more than 10% of the total contract value.

in good faith. There is, in any event, a possibility of the parties in a contract to incorporate a duty of good faith, a remedy when encountering unforeseen circumstances or when abrogating factors occur, such as misrepresentation, and this may be increasingly encountered in construction contracts as in the present one.

As observed by Lord Hobhouse in *Manifest Shipping v Polaris*[8]: "Having a contractual obligation of good faith in the performance of a contract presents no contractual term". It has also been observed by many prominent lawyers that the Doctrine of 'utmost good faith' is arguably a travesty of '*good faith*', in which it is used by parties as means to evade liabilities regardless of fairness. A duty of good faith has been implied, either by statute or by the courts. *The Unfair Terms in Consumer Contracts Regulations*, 1999 *Regulation* 5(1) refers to 'the requirement of good faith' in defining what an unfair term is. This compares with the continental legal systems (and even some other common law systems, particularly in the Unites States) where a general principle of good faith applies in contracts or, at least, in some aspects of contracts, and the doctrine of *force majeure* allows a degree of release or adjustment in the face of exceptional unforeseen conditions.

Under various laws as well, frustration sometimes provides relief where the performance or purpose of the contract has been made impossible or illegal by supervening events (i.e. events taking place after the contract was formed), which were not foreseen at time when the contract was done and which were not due to any act or emission of either party. The doctrine applies primarily to supervening events making performance impossible or illegal and that contractual obligations are no longer binding on the parties. The contract in this instance did not make any presentation for the problems that the Contractor has experienced during the execution of Project A such as the late staffing of the technical team at the start of the Project or lack of initial design which progressed into a large number of change orders. All the problems that were encountered and listed below could not be foreseen by both the Client and the Contractor and could be attributed to the doctrine of common mistake where the heading mistake refers to situations where one or both parties in a contract are under a misapprehension of present fact at the time of contract.

In summary to the aforementioned, the elements of the law which are pertinent to this particular global claim are misrepresentation, frustration and economic duress. The acts, preventions and the large number of change orders disputed, or in few of the cases were agreed and paid, provide the Contractor of Project A enough reasons under applicable laws to rescind this contract and if not so to substantially succeed in recuperating his losses under this statement of claim which amounts to a amount of USD 42,684,293.

The Contractor entered into a contract with the Client on 3 November 2006 to construct Project A for an amount of USD 61,987,185. It was obvious from the

---

[8] Manifest Shipping Co. Ltd. v Uni-Polaris Shipping Co. Ltd.: 'The Star Sea' (2001) 1 All ER 743.

early months of the contract that the conditions under which the Contractor thought it had tendered did not exist. The Contractor therefore submits that the contract could have been rescinded in accordance with the doctrine of applicable laws thereby limiting its losses in the early months of the contract. The Contractor contents that the contract could have been rescinded under at least one, if not all three, of the legal concepts listed above. Rescinding the contract for reasons of misrepresentation, frustration or economic duress would have allowed the Contractor to cut its losses in the early stages of the contract and to have realised an overall loss of a fraction of the losses it has currently incurred.

Rescinding the contract early would have minimised the Contractor's exposure to loss but it would have in turn exposed the Client to considerable additional costs. The project would have been re-tendered, or negotiated with the next lowest contractor, and the Client would have faced considerable delays and additional costs and losses. However, the Contractor valued its long-standing relationship with the Client, and its impeccable record in the Gulf region in the field of construction, and as a result it did not pursue the rescission of the contract. It did so trusting that an amicable settlement would be reached with the Client when the works were nearing completion.

8.8 Relief and Remedies in Law

The contract between the Client and the Contractor did not allow relief and remedies for these catastrophic and unprecedented events as they are usually areas where the law intervenes to provide relief or remedies despite apparently binding agreement between the parties to the contract. A representation in the contract is a statement of fact, past or present. A mere statement of opinion is not considered to be a representation, unless it implies the existence of past or present facts, for example, that reasonable grounds existed for the opinion or that there were no known contrary facts. A misrepresentation is a false representation. In particular, the expression is used to describe a statement made by one party, or its agent, to a prospective party to a contract so as to induce the contract. A representation may become incorporated as a term of the contract. If it is not so incorporated it is sometimes described as a mere representation.[9] A misstatement, on the other hand, may be a false statement of fact or it may be a piece of negligent advice. The expression is used to refer to liability under the rule in *Hedley Bryne v Heller*,[10] so a misstatement may induce a contract, but the liability of the maker of the statement does not depend on him being a party to the resulting contract.

---

[9] Heilbut Symons & Co v Buckleton (1913) AC 30.

[10] Hedley Byrne & Co Ltd **v** Heller & Partners Ltd (1964) AC 465.

The general and perhaps most difficult rule is that silence cannot amount to a representation. The general rule is subject to exception, where there is a duty to make disclosure. In this contract, the Client has not disclosed many essential points:

1. Site Restrictions.
2. Difficulties in mobilising good, experienced and technical professionals.
3. Unduly stringent safety and security procedures.
4. Unfinished studies.
5. Variations.

With construction contracts subject to the doctrine of *uberrimae fides* (utmost good faith), there is a positive duty on the parties to disclose known material facts. Failure to do so amounts to a misrepresentation entitling the contractor to rescind the contract. The test of materiality is whether the circumstances would influence the judgment of a prudent contractor in fixing the cost or in determining whether he will accept the risk. The right to rescind the contract depends not only on materiality, but also on whether the misrepresentation or non-disclosure included the making of the contract on the relevant terms.[11] The Contractor has entered this contract in utmost good faith and is claiming that the client has not disclosed a critical number of facts such as the relationship between the different parties and the structure of the Company which has proved an obstacle for the decision-making, the security restrictions on-site that became evident at the beginning of the project and the difficulty of obtaining decisions and information from the technical team.

In construction contracts, there is a continuing duty of disclosure after the contract has come into existence, but it is different from the pre-contract duty, both in its nature and the remedies available. This was not also the case as the Project evolved. In *Dillingham Construction Pty. Ltd. and Others v Downs,*[12] the Australian courts confirmed that in construction projects, the client had a duty to disclose information to tenderers. It was concluded that the essential question was that, in any particular case, the Client has generally an implied assumption of responsibility.

When a contract is subject to an express term amounting to a duty of good faith, this will include a duty to disclose material information such as stringent safety and security procedures. Although, depending on the precise nature of the term, this may not extend to a duty to have supplied such information prior to formation of the present contract, it will probably impose such a duty prior to an agreement of settlement or variation relating to the contract. Any such settlement or variation will, of course, be a fresh contract, but the duty of disclosure will arise from the original contract. There are further situations where non-disclosure may amount to misrepresentation.

---

[11]   Pan Atlantic Insurance Co. Ltd v Pine Top Insurance Co. Ltd (1994) 3 All ER 581.
[12]   (1972) 13 BLR 97.

Firstly, with any contract, where a positive representation has been made, failure to disclose subsequent changes which occur before the contract is made will amount to a misrepresentation.[13] In *With v O'Flangan*,[14] LJ Romer stated that:

> The only principle invoked by the appellants in this case is as follows. If A with a view to inducing B to enter into a contract makes a representation as to a material fact, then if at a later date and before the contract is actually entered into, owing to a change of circumstances, the representation then made would to the knowledge of A be untrue and B subsequently enters into the contract in ignorance of that change of circumstances and relying upon that representation, A cannot hold B to the bargain. There is ample authority for that statement and, indeed, I doubt myself whether any authority is necessary, it being, it seems to me, so obviously consistent with the plainest principles of equity.

Secondly, a partial representation, where non-disclosure of a fact distorts a positive representation, may constitute a misrepresentation.[15]

Thirdly, it is possible that active concealment might constitute misrepresentation. *BCCI v Ali*[16] was a test case brought to determine whether a party conduct was of sufficient gravity to be a breach of the duty of trust and confidence and, if so, what was the loss as a result of the breach and whether it should be compensated in damages. As per Lord Nicholls:

> In these circumstances there can be no question of BCCI having indulged in anything approaching sharp practice in this case. That being so, I prefer to leave discussion of the route by which the law provides a remedy where there has been sharp practice to a case where that issue arises for decision. That there is a remedy in such cases I do not for one moment doubt.

In Project A, the Client has treated the contract in such a way that all information required to construct as included and they have refused any further information. This created a clearly a situation of misrepresentation and non-disclosure where the Contractor has been subjected to much distress in executing the works and where further information needed to be disclosed.

8.9 The Global Claim Approach: Specific Factors

The global claim in this instance is a claim where the Contractor does not seek to attribute loss to specific breaches of the contract or to specific legal factors and doctrines that rendered this contract unworkable, but rather alleges a composite loss as a result of all the alleged breaches. A global claim is in this instance defined as "the antithesis of a claim where the causal nexus between the wrongful act or

---

[13] Ray v Sempers (1974) AC 370.

[14] (1936) Ch 575.

[15] Goldsmith v Rodger (1962) 2 Lloyd's Rep 249.

[16] Bank of Credit and Commerce International SA v. Munawar Ali, Sultana Runi Khan and Others (2001) 1 All ER 961.

omission of the defendant and the loss of the claimant has been clearly and intelligibly pleaded."[17]

This global claim is manifested in this case by not reconstructing what actually happened on-site as it is impossible or impractical to plead and prove the *causal nexus*. The factors and reasons that the Contractor has incurred are so intertvened that he could not possibly identify how each breach has caused certain damage and therefore an amount of loss. Therefore, the contractor is pleading his case using a global claim as global claims are only permissible when it is impossible or impractical to plead and prove the *causal nexus*. See *J. Crosby & Sons Ltd v Portland Urban District Council*[18] and *London Borough of Merton v Stanley Hugh Leach.*[19]

The composite loss in this claim is prepared as a modified total cost method,[20] where the quantification of loss is achieved by subtracting the tender cost of the works from the final cost. Then, this total cost is further purified by subtracting from it the direct and indirect costs and then by subtracting all the variations that have a direct cost impact and the material escalation factor. The variations with a direct cost impact and the material escalation will be claimed separately under different claims.

In this claim, the Contractor argues that the events which occurred during this contractual period were complicated as such to make it impracticable, if not impossible, to assess the additional expenses caused by delay and disorganisation due to any one of the events in isolation from the other events. While the Contractor may be able to provide a list of numerous events which have caused disruption to his works, the global claim does not prove what the effects of such disruptive events really were to the works as they cannot be defined in terms of *quantum*. It can be argued that there can be many reasons why the Contractor's final costs are more than his tendered costs as he may simply have tendered too low in the first place. In this case, the Contractor won the contract through a tender process while competing with a number of other bidders. Further negotiation were conducted between the Client and the Contractor, after the tender process has ended, which resulted in a further reduction to his bid and therefore the cost of his works cannot be argued as too low.

Another argument the Client can arise is that a global claim makes the huge assumption that a'l the additional time and costs were caused by the disruptive events which are included for the purpose of this claim. The Contractor is claiming justifiably that all the events that caused the disruption to the works were the Client's responsibility and to act in a fair and reasonable manner has deducted from his calculation all direct and indirect costs as well as the profit margin. The loss the Contractor is claiming for is only the actual losses to the works he had to

---

[17]   Bernhard's Rugby Landscapes Ltd v Stockley Park Consortium Ltd (1997) 82 BLR 39.

[18]   (1967) 5 B.L.R. 121.

[19]   (1985) 32 BLR 68.

[20]   Servidone Construction Corporation v the United States 931 F.2d 860. April 24, (1991).

carry out over and above his original scope of work. It is also to be noted that the Contractor has accelerated the works to finish the works by the agreed completion date as instructed by the Client.

The global claim pursued is usually based on an allegation that there were numerous variations in the contract and the cost overran. The claimant here alleges that the cost overrun is recoverable as a result of the variations and the Client initiated causative factors such a lack of design, late approval of the design and materials submittals and vitiating law doctrines that rendered the Contract frustrated. There is, however, no analysis that a particular causative factor lead to a particular item of loss. This global claim is based on the principle that there were numerous events interfering with the works entitling the Contractor for damages and incurred losses. Again there is no link between the alleged events and damages but the Contractor is claiming that the disruptive effects together with the legal deviation from the Contract have contributed as whole to a lump sum loss of USD 42,684,293.

The law states that if parts of the claim can be pleaded and proved on a conventional basis and parts cannot, only the latter can and should be prosecuted on a global basis. In this case, the claimant in his modified total cost approach has eliminated from his global loss all losses that could be claimed conventionally such as:

1. Material escalation due to the unrealistic and non-envisaged increases that were recorded in the prices of all construction materials mainly from 2006 to 2010.
2. Variations that have a direct cost impact and could have a monetary value.

Having this claim submitted as a global claim was decided due to the following factors:

- There is no need to do a detailed analysis of the variations and events on-site. This was important due to the lack or the few identified factual supporting documents as most of the communication was done on 'good will' basis due to the long-standing relationship between the Client and the Contractor. A global claim in this instance is permissible where it is impractical to disentangle that part of the loss attributable to each head of claims shown above, and the situation as in Project A, has not been brought about by delay or other conduct on the part of the Contractor. In such circumstances the court infers that the defendant breaches caused the extra cost or cost overrun and the *causal nexus* was inferred rather than demonstrated.
- It is more efficient and precise to present and defend. This is critical as the Contractor needs to finalise this claim as quick as he can due to the depilating losses he incurred which amounts to more than three times the amount of liquidated damages or the penalties he could have incurred if he terminated this contract. It is also important in case this claim goes for arbitration or to the courts where the fundamental concern of the court is that the dispute between the parties should be determined expeditiously and economically and, above all, fairly. While a plaintiff is entitled to present its claim as he thinks fit,

a defendant is entitled to know the case which he has to meet with as much certainty and particularity as is reasonable, having regard to the circumstances and to the nature of the acts themselves by which the damage is done.[21] In this case and with the circumstances and facts present, the Contractor is presenting this claim as a global claim to settle this matter expeditiously.

- The Contractor realises that the correct manner of presenting a claim is to link the cause with the effect. However, in this case, this is not achievable, especially when the particulars of this global claim, rather than a prolongation claim, are disruption events with no methodology or a systematic approach to enable their calculations. To counter such difficulties, the Contractor has proved that to link cause and effect is not possible and herewith the use of this global claim.

Another important issue has arisen in which the supporting documents establishing the matrix of facts have proven to be of such quality and number that it assisted the Contractor in writing a claim but proved impossible to establish the *quantum* for each causative effect. It also took such a prohibitive long time for the Contractor to collect statement of facts and documents and to try to find complex and time-wasting procedures for documentation and calculation methodologies that it jeopardised the completion of this claim.

8.10   Effect of Variations and Disputed Variations

The following table lists the claim events which have been disputed throughout the contract period for Project A.[22] This list is divided into two parts where Part 1 includes all variations that constitute part of the global claim. These variations are described in detail as follows:

- To list all correspondence related to this claim.
- To describe the problem and to relate each correspondence to the factual nexus of the case which in this case the variation and the dispute that had occurred.
- To identify the contractual clauses relevant to this variations. For this type of sub-claim, the main contract clauses which govern the arguments presented under this head of claim are Clause 2 (Contractor Obligations), Clause 10 (Changes) and Clause 19 (Claims Settlement; Disputes). To establish liability on the Client by highlighting the Client reluctance to issue the change orders.
- To establish the liability in between the variation and the relevant contractual clauses.

---

[21]   Ratcliffe v Evans (1892) 2QB 524.

[22]   For the purpose of the book only a limited number of variations are provided to give the reader a flavour of the issues that can arise in a particular project and for him to be able understand how to structure these types of sub-claims.

Table 8.2 List of variations

Claim Nr.	Head of claim	Amount in USD
Variations—part of the global claim		
1	Additional design for electro-practical works	No figure part of global claim
2	Provision of conduits and boxes for the translation system in the auditorium	No figure part of global claim
3	Demolition work at B120—Zone 3B	No figure part of global claim
4	Removal of concrete blocks in the vicinity of B159	No figure part of global claim
5	Omission of the structural design affecting the execution of finishing works—B100	No figure part of global claim
6	Dummy columns pedestal design changes at B100 Roof	No figure part of global claim
Variations—agreed		
7	Pigmented epoxy coating for communications room Flooring—B100	63,450 USD
8	Energising project relay setting at RMUs 346 & 346A	44,870 USD
9	Energising project relay setting at RMUs 345 & 345A	22,560 USD
10	Changing lighting installation details at B100 auditorium entrance	17,670 USD
11	Changing door leaves from Type B to Type A & Type C at B100	8,700 USD
12	Changing doors from non-fire rated to fire rated Doors	34,760 USD
13	Change steel door frames to wood frames and wood architraves	23,360 USD
14	Provision of openings at roof of AQMS	8,820 USD
15	Closing gap between curtain walling and pre-cast panels at B100 main entrance	85,880 USD
16	Provision of additional acoustic false ceiling at B100—Zone D	22,100 USD
17	Additional pedestrian rurnstile at B105	62,700 USD
Total amount—agreed variations		**394,870 USD**

As explained earlier in this Chapter, these variations are so intervened where an individual value cannot be identified as this part of the work so intervened with other activities as it is not possible to give it a monetary value.

Part 2 includes variations that can be quantified and therefore were omitted from the amount claimable in this claim and to be claimed as a separate claim. For the purpose of this book and for not being too repetitious it was decided that, for the variations in this part, a list of correspondences and variation value are only included (Table 8.2).

Table 8.3 List of
correspondence—Additional
design for electro-practical
works

Date	Reference	Sender
19 March 2009	IS/L0102-09	The contractor
28 March 2009	PMC 394-0147-09	The client
11 April 2009	IS/L0149-09	The contractor
23 April 2009	PMC 394-0233-09	The client
29 April 2009	IS/L0179-09	The contractor
30 April 2009	PMC 394-0248-09	The client
26 May 2009	PMC 394-0314-09	The client
28 May 2009	IS/L0227-09	The contractor
01 January 2009	IS/L0285-09	The contractor
07 July 2009	PMC 394-0433-09	The client

8.11   Variations: Part of the Global Claim

8.11.1   Additional Design for Electro-Practical Works

8.11.1.1   List of Correspondence

Under this heading the following correspondence has been received (Table 8.3).

8.11.2   Analysis

Considerable correspondence was exchanged on this subject between March 2009 and July 2009. In his letter reference IS/L0102-09 dated 19 March 2009 the Contractor explained that the contract drawings did not incorporate any provision for exhaust fans to be connected to the BMS system and for other essential electrical and practical components for the Project. Consequently, pursuant to the express provisions of Clause 10 of the general conditions of contract, the Contractor requested that a change order be issued to cover this additional work if the Client wants to instruct the Contractor to prepare the design works.

Subsequent memoranda from the Client refuted this position and suggested that the subject works were referred to in the technical specifications and were inferable from the contract documents generally. In his letter reference IS/L0149-097 dated 11 April 2009, the Contractor responded and made the clear point that the technical specifications required the electrical drawings to be followed to determine the express requirement for the necessary connections.

The Client issued a letter reference PMC 394-0248-09 dated 30 April 2009 which requested the Contractor to attend a meeting convened to resolve the impasse over this dispute. The Contractor attended the meeting and was under the impression that the issue was resolved and that the Client would issue subsequent letter to officially instruct the Contractor to execute the additional work.

This instruction arrived in the form of a memorandum from the Client (PMC-394-0314-09 dated 26 May 2009); however, this memorandum instructed the Contractor to submit an official offer for additional design works but fell short of issuing the required change order. In his letter reference IS/L0227-09 dated 28 May 2009 the Contractor responded to make the point that the Client ignored the agreement reached during the meeting. However, in this letter the Contractor confirmed that he would proceed to execute the additional work but made it clear that the instructed work was additional to the scope of the contract and that the Contractor intended to claim for all the related costs.

8.11.3 Contractual Liability

The contract clauses which govern the arguments presented under this head of claim are Clause 2 (Contractor Obligations), Clause 10 (Changes) and Clause 19 (Claims Settlement; Disputes). Clause 2 (Contractor Obligations) establishes and defines the scope of the Contractor obligations in terms of this lump sum contract. This clause explains that the scope of the works as shown on the contract drawings and included in the contract specification form the basis of the contract price irrespective of what work is measured in the bill of quantity. Clause 10 (Changes) establishes the rules regarding the process of changing the scope of the works and this clause restricts additional works-to-works covered by a specific change order. Clause 19 (Claims Settlement; Disputes) establishes the rules with respect to claims the Contractor wishes to make to obtain compensation for executing additional work or other loss and expense not envisaged in the original scope of works or the contract documents.

In this instance the Contractor has been denied his contractual entitlements pursuant to the express provisions of Clause 10 and has no alternative but to seek compensation through the express provisions of Clause 19.

The MCCs for external exhaust fans, required to be connected to the BMS, were neither included on the electrical drawings nor included in the contract bill of quantity. The general specification made reference to a requirement for these MCCs but clearly stated that the electrical drawings were to be followed for the specific requirement. The Client asked the Contractor to design these details but the Contractor considered these to be outside the scope of work.

Pricing lump sum contracts places considerable risk on the Contractor. Design do not form part of the contract and this places a considerable risk with the Contractor. To be accurate the Contractor must ensure that required design work required by him reflects the true scope of work and to be competitive the Contractor must ensure that no more than the maximum scope of works is reflected in the tender return. Including costs for items which are outside the Contractor's scope of work would make the tender uncompetitive and ultimately result in the Contractor's tender being unsuccessful.

Table 8.4 List of correspondence—Provision of conduits and boxes for the translation system in the auditorium

Date	Reference	Sender
10 June 2007	IS/L0249-07	The contractor
20 November 2007	PMC 394-0744-07	The client
15 December 2007	IS/L0415-07	The contractor
02 January 2008	IS/L0004-08	The contractor
08 January 2008	PMC 394-0013-08	The client
09 January 2008	PMC 394-0024-08	The client
10 January 2008	PMC 394-0027-08	The client
26 January 2008	PMC 394-0069-08	The client
12 February 2008	IS/L0060-08	The contractor
25 February 2008	PMC 394-0180-08	The client
01 April 2008	IS/L0113-08	The contractor
24 April 2008	PMC 394-0419-08	The client

In this particular instance the Contractor rightly considered the MCCs, which were not included on the contract drawings, and for him to redesign this part of the work was outside the original scope of works.

8.11.4 Provision of Conduits and Boxes for the Translation System in the Auditorium

8.11.4.1 List of Correspondence

Under this heading the following correspondence has been received (Table 8.4).

8.11.5 Analysis

Considerable correspondence was exchanged on this head of claim yet it is a relatively straight forward issue. The translation system in the auditorium was designed without the provision of a system of conduits and boxes to accommodate the communication cables. Presumably the designers had in mind that the Company would have installed a wireless audiovisual system. However, during a meeting in May 2007 the Contractor was informed that the design of the audio-visual system would require the installation of a system of conduits and boxes. The Contractor requested confirmation but did not receive this confirmation from the Client.

In the interim, the Contractor continued with the execution of the concrete works and did not install the system of conduits and boxes. Under the conditions of contract the Contractor must receive an official change order to facilitate the commencement of additional works. The Client did not contact the Contractor again until the end of November 2007 to remind the Contractor that he did not follow the instructions issued during the meeting held in May 2007.

Installing the requested system of conduits and boxes now involved the additional effort of cutting the slabs and installing the system which was more expensive to execute. The remaining correspondence generally deals with the details of the installation of the system and does not contain any suggestion that the Company would not issue a change order; however, to the completion date a change order has not been issued.

The Contractor issued his initial letter reference IS/L0249-07 dated 10 June 2007 requesting confirmation of the instruction to install a system of conduits and boxes translation system in the auditorium and requesting a change order to be issued. The Client did not respond until 20 November 2007 via their letter reference PMC 394-0744-07 wherein he queried why the concrete had been poured without installing the necessary conduits and boxes. This letter also requested the Contractor to submit proposals to retrofit the system. The Contractor responded via letter IS/L0415-07 dated 15 December 2007 and reminded the Client that their letter in June offered two proposals and requested a change order which the Company never issued. The Contractor also justified why he progressed with the works without installing the system. Subsequent correspondence exchanged to the end of January 2008 deals with technical issues related to the installation of the system. On 12 February 2008 via letter reference IS/L0060-08 the Contractor again requested an official change order to cover the additional works. The client responded via their letter reference PMC 394-0180-08 dated 25 February 2008 and attached a memorandum from the engineer which stated that the change order will be processed which was never done.

8.11.6 Contractual Liability

The contract clauses which govern the arguments presented under this head of claim are Clause 2 (Contractor Obligations), Clause 10 (Changes) and Clause 19 (Claims Settlement; Disputes). Clause 2 (Contractor Obligations) establishes and defines the scope of the Contractor's obligations in terms of this lump sum contract. This clause explains that the scope of the works as shown on the contract drawings and included in the contract specification form the basis of the contract price irrespective of what work is measured in the bill of quantity. Clause 10 (Changes) establishes the rules regarding the process of changing the scope of the works and this clause restricts additional works-to-works covered by a specific change order. Clause 19 (Claims Settlement; Disputes) establishes the rules with respect to claims the Contractor wishes to make to obtain compensation for executing additional work or other loss and expense not envisaged in the original scope of works or the contract documents.

Table 8.5 List of
correspondence—Demolition
work at B120—Zone 3B

Date	Reference	Sender
01 May 2009	PMC 394-0254-09	The client
16 May 2009	IS/L0203-09	The contractor
05 May 2009	PMC 394-0256-09	The client
22 May 2009	IS/L0216-09	The contractor
30 May 2009	PMC 394-0334-09	The client

In this instance the Contractor has been denied his contractual entitlements pursuant to the express provisions of Clause 10 and has no alternative but to seek compensation through the express provisions of Clause 19.

It would appear that the Company is not refuting the validity of this additional work; however, for the official change order was not issued. The additional work of installing a system of conduits and boxes translation system in the auditorium should have been instructed pursuant to the express wording of Clause 10 and evaluated accordingly. The Contractor decided, in the best interest of completing the works, to execute the additional works and seek compensation through the express provisions of Clause 19 of the General Conditions of Contract.

8.11.7 Demolition Work at B120: Zone 3B

8.11.7.1 List of Correspondence

Under this heading the following correspondence has been received (Table 8.5).

8.11.8 Analysis

Building No. B120 in Zone 3B was to be demolished and removed as part of the contract documents including the bill of quantity (item reference 020-60-7-2). When it became apparent on-site that the building was in fact a steel prefabricated building the Client decided that it would have a residual value and consequently requested the Contractor to carefully dismantle the building and erect same in a new location that would be advantageous to the Client. The Contractor informed the Client that the provision of the contract stipulated that all materials salvaged from demolition activities was the property of the Contractor. Consequently, the Contractor requested a variation order to reimburse the costs of carefully dismantling the structure, erecting the structure in a new location and for compensation for the salvage materials. The Client disputed this.

Table 8.6 List of correspondence—Removal of concrete blocks in the vicinity of B159

Date	Reference	Sender
22 July 2007	PMC 394-0473-07	The client
23 July 2007	IS/L0040-07	The contractor
31 October 2007	IS/L0089-07	The contractor
17 February 2008	IS/L0064-08	The contractor
26 February 2008	PMC 394-0185-08	The client
02 March 2008	IS/L0086-08	The contractor
27 April 2008	PMC 394-0412-08	The client
03 May 08	IS/L0111-08	The contractor
15 May 2008	PMC 394-0487-08	The client
26 May 2008	IS/L0144-08	The contractor

8.11.9 Contractual Liability

The main contract clauses which govern the arguments presented under this head of claim are Clause 2 (Contractor Obligations), Clause 10 (Changes) and Clause 19 (Claims Settlement; Disputes). In this instance the Contractor has been denied his contractual entitlements pursuant to the express provisions of Clause 10 and has no alternative but to seek compensation through the express provisions of Clause 19.

This was a variation to the original scope of works. The contract documents clearly made provision for the value of the salvaged material to transfer to the Contractor. The Client should have made a fair assessment of the value of the materials salvage and the costs of carefully dismantling and erecting the structure in a new location.

8.11.10 Removal of Concrete Blocks in the Vicinity of B159

8.11.10.1 List of Correspondence

Under this heading the following correspondence has been received (Table 8.6).

8.11.11 Analysis

This head of claim relates to an argument over the correct or most appropriate definition of very large concrete blocks which the Company instructed the Contractor to remove from the site. The Contractor removed approximately 48 large concrete blocks ranging in size from 0.6 m x 0.6 m x 0.6 m (8 Nr.) to 2.0 m x 1.6 m x 0.7 m (40 Nr.). The removal of these blocks was not included on the drawings and no specific reference was made in the contract specification. In an attempt to avoid incurring additional cost the Company decided to classify the removal of these blocks as '*clearing and grubbing*' while the Contractor insists that this is additional demolition works and has requested appropriate compensation.

The exchange of correspondence commenced on 22 July 2007 when the Client issued their letter reference PMC 394-0473-07 and therein the Client initial instruction to remove the blocks as debris from the site. The next day the Contractor responded via their letter reference IS/L0040-07 dated 23 July 2007 and notified that the blocks would be removed and the additional cost calculated and submitted to the Client. The Contractor also informed the Client that there were 48 large concrete blocks to be removed. The Client did not respond to this letter and the Contractor considered that the Company accepted its position.

The Contractor issued a further letter, IS/L0064-08, dated 17 February 2008 and informed the Company that it was removing the concrete blocks and requested a change order to cover the additional works. This letter also attached a memorandum which outlined the effort involved. To this letter the Client responded via their letter reference PMC 394-0185-08 dated 26 February 2008 and stated that the blocks were debris and must be removed under clearing and grubbing at no additional cost. The Contractor responded via letter reference IS/L0086-08 dated 02 March 2008 and informed the client that the very large concrete blocks could not be considered clearing and grubbing and could only be considered demolition and removal works which were excluded from the scope of the work included in the contract. The Contractor reiterated its request that the Company issue an appropriate change order.

The Client responded via their letter reference PMC 394-0412-08 dated 27 April 2008 and informed the Contractor to submit photographic records to justify its position. The Contractor replied via their letter reference IS/L0111-08 dated 03 May 2008 and informed the Client that it was impossible to provide photographic records as the blocks had already been broken up and removed from site. The Contractor went further to record that the Client had been advised of the magnitude of the blocks in July 2007 and it was then that the Client should have inspected the problem and requested whatever records the believed were necessary. The Client responded via their letter reference PMC 394-0487-08 dated 15 May 2008 and stated that additional costs would not be considered. The Contractor responded and notified the dispute pursuant to the express provisions of Clause 10.5 of the General Conditions.

8.11.12 Contractual Liability

In this instance the Contractor has been denied his contractual entitlements pursuant to the express provisions of Clause 10 and has no alternative but to seek compensation through the express provisions of Clause 19.

This head of claim could be construed as an assessment of whether the large concrete blocks are to be considered demolition and removal works or clearing and

Table 8.7 List of correspondence—Omission of structural design affecting execution of finishing works—B100

Date	Reference	Sender
12 September 2009	IS/L0203-09	The contractor
01 October 2009	PMC 394-0639-09	The client
11 October 2009	IS/L0207-09	The contractor
21 October 2009	PMC 394-0656-09	The client

grubbing. However, irrespective of this assessment the fact remains that the Contractor informed the Client in July 2007 that it considered the removal of the blocks to be additional to the contract and the Client did not reply until February 2008 when the Contractor requested a change order. Apart from the fact that the Contractor is correct in its assessment that the breaking up and removal of the very large concrete blocks is rightly considered demolition as opposed to clearing and grubbing; the Client inaction from July 2007 is considered as acceptance of the contractor argument.

The additional work of demolishing and removing the large concrete blocks should have been instructed pursuant to the express wording of Clause 10 and evaluated accordingly. The Contractor has no alternative but to seek compensation through the express provisions of Clause 19 of the General Conditions of Contract.

8.11.13 Omission of Structural Design Affecting Execution of Finishing Works: B100

8.11.13.1 List of Correspondence

Under this heading the following correspondence has been received (Table 8.7).

8.11.14 Analysis

The Contractor notified the Client of a design discrepancy through their letter reference IS/L0203-09 and IS/L0207-09 dated 12 September 2009 and 11 October 2009. The letters explained that there was a difference in level between the double slabs located at B100—fourth floor Zone B Axis 8—202/G-J. The Contractor informed the Client that this discrepancy was affecting several work items such as stainless cladding, stainless steel windows and the false ceiling in the affected locations. The Contractor proposed a frame solution which was attached to the above referenced letter and advised the Client that this additional work would need to be instructed pursuant to Clause 10 of the Contract to facilitate the execution of the additional works. However, the Contractor also informed the Client that because the additional works needed urgent approval that the Contractor would

Table 8.8 List of correspondence—Dummy columns pedestal design changes at B100 roof

Date	Reference	Sender
16 January 2008	IS/L0015-08	The contractor
02 February 2008	PMC 394-0092-08	The client
11 February 2008	IS/L0058-08	The contractor
09 March 2008	PMC 394-0212-08	The client
13 March 2008	IS/L0095-08	The contractor
23 March 2008	PMC 394-0267-08	The client
25 March 2008	IS/L0108-08	The contractor

proceed to execute the additional works if the Company did not approve the change order within three days.

The Client responded via their letter reference PM 394-0639-09 dated 01 October 2009 and PMC 394-0656-09 dated 21 October 2009. They stated that the difference in level between the double tee slabs was a design intention and that the frame required to resolve the level differential was also part of the original design. As such the Client considered that the frame would not be considered additional to the original scope of works.

8.11.15 Contractual Liability

The General Conditions of Contract make express provision to identify discrepancies in the original contract documents and have these clarified and resolved by the Company. The Contractor contends that the frame required to correct the apparent level differential in the slabs was not included in the original scope of work and that the Company was required to consider this as additional to the scope and issue an appropriate change order pursuant to the express requirements of Clause 10.

In this particular instance the Contractor rightly considered the steel frame, which was not included on the contract drawings, to be outside the original scope of works and tendered accordingly. The additional work of adding this frame should have been instructed under the express wording of Clause 10 and evaluated accordingly. The Contractor decided, in the best interest of completing the works, to execute the additional works and seek compensation through the express provisions of Clause 19 of the General Conditions of Contract.

8.11.16 Dummy Columns Pedestal Design Changes at B100 Roof

8.11.16.1 List of Correspondence

Under this heading the following correspondence has been received (Table 8.8).

8.11.17 Analysis

This head of claim is explained in detail through the above referenced correspondence issued between the 16 January 2008 and the 25 March 2008. In essence
the Contractor had previously submitted shop drawings for approval for six
number pedestal bases on the fifth floor roof of building B100 and the shop
drawings were approved. The shop drawings were submitted to reflect the size of
the pedestal bases that were clearly designed and shown on the contract drawings.
During the execution of the work the Contractor discovered that the size of the
pedestal bases was not appropriate in terms of supporting the dummy columns that
were required. To make the correction the Contractor issued a *'field adjustment'*
request on 11 December 2007 and followed this field adjustment request with letter
reference IS/L0015-08 dated 16 January 2008.

The Company responded via their letter reference PMC 394-0092-08 dated
02 February 2008 and stated that a change order was not appropriate because the
Contractor field adjustment did not confirm with the scope of work requirements. The
Contractor responded via letter IS/L0058-08 dated 11 February 2008 acknowledging
their original mistake of including the additional works on the wrong field adjustment
form—the additional works were inadvertently included on the *'Subcontractor
Adjustment'* form by mistake. However, in this letter the Contractor requested the
Company to ignore the clerical error and issue the required change order.

The Client responded via their letter reference PMC 394-0212-08 dated
09 March 2008 and stated that determining the size of the pedestals was the
responsibility of the Contractor because the columns were decorative columns.
The Contractor responded via his letter IS/L0095-08 dated 13 March 2008 and
clearly stated that the design of the pedestals rested with the Company and their
original shop drawings submitted for approval were based on this design criteria.
The design criteria subsequently proved to be incorrect and the field adjustment
was required to make the correction.

A subsequent letter from the Client reiterated their previously stated position and a
follow-up letter from the Contractor requested the issue to be taken to dispute level.

8.11.18 Contractual Liability

Under this contract the design risk has been retained fully by the Company and has
not been transferred in any capacity to the Contractor. The original size of the
pedestal bases on the roof of building B100 was not correct and this was discovered by the Contractor during the execution of the works. The additional work
of adjusting the size of the concrete bases should have been instructed pursuant the
express wording of Clause 10 and evaluated accordingly. The Contractor decided,
in the best interest of completing the works, to execute the additional works and
seek compensation through the express provisions of Clause 19 of the General
Conditions of Contract.

8.12 Variations: Agreed

8.12.1 Pigmented Epoxy Coating for Communications Room Flooring: B 100

8.12.1.1 List of Correspondence

This variation was valued at 63,450 USD (Table 8.9).

8.12.2 Energising Project Relay Setting at RMUs 346 & 346A

8.12.2.1 List of Correspondence

This variation was valued at 44,870 USD (Table 8.10).

8.12.3 Energising Project Relay Setting at RMUs 345 & 345A

8.12.3.1 List of Correspondence

This variation was valued at 22,560 USD (Table 8.11).

8.12.4 Changing Lighting Installation Details at B100 Auditorium Entrance

8.12.4.1 List Correspondence

This variation was valued at 17,670 USD (Table 8.12).

8.12.5 Changing Door Leaves from Type B to Type A & Type C at B100

8.12.5.1 List Correspondence

This variation was valued at 8,700 USD (Table 8.13).

Table 8.9 List of correspondence—Pigmented epoxy coating for communications room flooring—B 100

Date	Reference	Sender
27 November 2007	PMC 394-0774-09	The client
13 November 2009	IS/L0384-09	The contractor
11 November 2009	IS/L0381-09	The contractor

Table 8.10 List of correspondence—Energising project relay setting at RMUs 346 & 346A

Date	Reference	Sender
12 November 2008	PMC 394-0386-09	The client
14 August 2009	PMC 394-0518-09	The client
26 August 2009	IS/L0335-09	The contractor
01 September 2009	PMC 394-0569-09	The client
04 September 2009	IS/L0340-09	The contractor
05 September 2009	PMC 394-0586-09	The client
29 September 2009	IS/L0355-09	The contractor
21 November 2009	PMC 394-0748-09	The client
03 December 2009	IS/L0407-09	The contractor

Table 8.11 List of correspondence—Energising project relay setting at RMUs 345 & 345A

Date	Reference	Sender
03 April 2008	PMC 394-0313-08	The client
02 April 2008	IS/L 0117-08	The contractor

Table 8.12 List of correspondence—Changing lighting installation details at B100 auditorium entrance

Date	Reference	Sender
22 April 2008	PMC 394-0393-08	The client
06 January 2008	IS/L0011-08	The contractor
03 June 2008	PMC 394-0559-08	The client

Table 8.13 List of correspondence—Changing door leaves from type B to type A & type C at B100

Date	Reference	Sender
01 February 2008	IS/L0020-08	The contractor
11 March 2008	PMC394-0234-08	The client
24 April 2008	PMC 394-0392-08	The client

8.12.6 Changing Doors from Non-Fire Rated to Fire Rated Doors

8.12.6.1 List of Correspondence

This variation was valued at 34,760 USD (Table 8.14).

Table 8.14 List of
correspondence—Changing
doors from non-fire rated to
fire rated doors

Date	Reference	Sender
14 January 2008	IS/L0013-08	The contractor
22 April 2008	PMC 394-0393-08	The client
18 May 2008	IS/L0175-08	The contractor
03 June 2008	PMC 394-0559-08	The client

Table 8.15 List of
correspondence—Change
steel door frames to wood
frames and wood architraves

Date	Reference	Sender
18 January 2008	IS/L0019-08	The contractor
04 February 2008	PMC 394-0095-08	The client
11 February 2008	IS/L0056-08	The contractor
19 February 2008	PMC 394-0155-08	The client
23 February 2008	IS/L067-08	The contractor

Table 8.16 List of
correspondence—Provision
of openings at roof of AQMS

Date	Reference	Sender
06 January 2008	PMC 394-006-08	The client
26 January 2008	IS/L0032-08	The contractor
06 February 2008	PMC 0114-08	The client
31 March 2008	PMC 0303-08	The client
05 April 2008	IS/L0118-08	The contractor
27 April 2008	PMC 0414-08	The client

8.12.7 Change Steel Door Frames to Wood Frames and Wood Architraves

8.12.7.1 List of Correspondence

This variation was valued at 23,360 USD (Table 8.15).

8.12.8 Provision of Openings at Roof of AQMS

8.12.8.1 List of Correspondence

This variation was valued at 8,820 USD (Table 8.16).

Table 8.17 List of correspondence—Closing gap between curtain walling and pre-cast panels at B100 main entrance

Date	Reference	Sender
17 February 2008	IS/L0062-08	The contractor
26 February 2008	PMC 0188-08	The client
02 March 2008	IS/L0084-08	The contractor
29 March 2008	PMC-SO-394-0294-08	The client
05 April 2008	IS/L0119-08	The contractor
27 March 2008	PM 394-0397-08	The client
03 May 2008	IS/L0153-08	The contractor
13 May 2008	PMC 0481-08	The client

Table 8.18 List of correspondence—Provision of additional acoustic false ceiling at B100

Date	Reference	Sender
17 February 2008	IS/L0061-08	The contractor
26 February 2008	PMC 0187-08	The client
02 February 2008	IS/L0082-08	The contractor
23 March 2008	PMC 0273-08	The client
25 March 2008	IS/L0105-08	The contractor
17 April 2008	PMC 0355-08	The client

Table 8.19 List of correspondence—Additional pedestrian turnstile at B105

Date	Reference	Sender
24 May 2008	IS/L0141-08	The contractor
03 June 2008	PMC-SO-394-0562-08	The client

8.12.9 Closing Gap Between Curtain Walling and Pre-Cast Panels at B100 Main Entrance

8.12.9.1 List of Correspondence

This variation was valued at 85,880 USD (Table 8.17).

8.12.10 Provision of Additional Acoustic False Ceiling at B100

8.12.10.1 List of Correspondence

This variation was valued at 22,100 USD (Table 8.18).

8.12.11 Additional Pedestrian Turnstile at B105

8.12.11.1 List of Correspondence

This variation was valued at 62,700 USD (Table 8.19).

8.13   Contract Provisions: General Terms and Conditions

In this Contract the following words and expressions shall have the following meanings assigned to them:

'Company Representative' means a party duly authorised by Company to act on behalf of Company to administer the Contract on its behalf, with whom Contractor may consult at all reasonable times, and whose instructions, request and decisions shall to the extent stipulated in the Contract or delegated by Company be binding on Company.

'Engineer' means the design and supervising contractor responsible for the technical supervision and inspection of the Work and for clarifying the design and/or providing supplementary design, with duties and authorities as described for him in the Contract documents or as may be delegated by Company/Company Representative. Engineer shall be forwarded through the Company Representative and when issued accordingly shall have the same effect as given by the Company Representative.

'Contractor Representative' means a party or parties duly authorised by Contractor to act on behalf of Contractor, with whom Company may consult at all reasonable times, and whose instructions, requests, and decisions shall be binding on Contractor as to all matters pertaining to this Contract.

All dates and periods shall refer to the Gregorian calendar. 'Day' means a calendar day and 'year' means 365 days unless a leap year having 366 days (leap years are 2004 and 2008).

'Effective Date' means the date of acceptance of the Letter of Intent.

'Work Commencement Date' is the date stated in the Notice to proceed as the date on which Contractor shall commence the Work.

The 'Time for Completion' means the agreed date set forth in the Form of Agreement upon which Contractor shall achieve Practical Completion as described later in this Contract.

'Project Completion' shall be achieved when all Exception Items have been completed or corrected by Contractor.

'Critical Milestone Dates' means the agreed dates by which specified portions of the Work are to be completed.

'Work Site' or Site means all locations at which Contractor performs any portion of the Work.

'Work' means any and all temporary and permanent works and services to be carried out by Contractor under this Contract and includes Materials as appropriate.

'Materials' means materials, equipment, machinery, apparatus to be constructed, modified or provided by Contractor under this Contract.

'Contractor's Equipment' means support facilities, machinery, construction plant, equipment, tools accessories, spare parts, appliances and things of every kind required in carrying out and completing the Work.

'Contractor's Personnel' mean all personnel provided by Contractor hired or otherwise assigned by Contractor including those of subcontractors to carry out labour and supervision required to execute the Work.

'Change Order' means any written alteration of this Contract expressly designated as a Change Order and signed by both parties. All Change Orders must be signed on behalf of Company, by the Contract signatory or the incumbent of the Contract signatory's position.

'Documents' means all drawings, specifications, design and engineering documents, calculations, samples, patterns, models, operation and maintenance manuals and other documents to be submitted by Contractor to Company under this Contract, regardless of the forms or media in or on which such Contractor's products may exist.

8.14   Contract Provisions: Contractor Obligations

Contractor shall perform the Work in a diligent, efficient and workmanlike manner with reasonable promptness and dispatch, in accordance with Company's instructions and its work rules and regulations in regard to safety, efficiency and good conduct of workmen.

Contractor shall, at his own cost and responsibility, perform all the works and services required for design (as specified in the Scope of Work), procurement of materials and equipment, manufacture, fabrication, delivery, construction and/or modification, erection, installation, commissioning, testing and remedying of any defects of the Work, but including any and all works, services, surveys, coordination, warranties, temporary work, provisions, arrangements and the like, which may be required of Contractor to do for complete accomplishment of the intent of this Contract all strictly in accordance with the specifications and requirements set out in the Contract as well as the applicable standards generally acknowledged in the industry.

In the event that the Contractor finds any ambiguity, discrepancy or omission in the specifications and/or requirements set out in the Contract, Contractor shall immediately notify Company of the same in writing. Company shall interpret, clarify, adjust or make necessary determination and shall promptly issue written instructions directly in what manner the Work shall be carried out. Any adjustment made by Contractor without Company's written approval shall be at Contractor's risk and expense.

This is a lump sum contract. Quantities in the Bill of Quantity.

The quantities set out in the Bill of Quantity are the estimated quantities for the Work and they are not to be taken as the actual and correct quantities of the Work. Contractor shall be responsible for verifying and checking all the quantities. If work is shown in the Scope of Work, on Drawings, or described in the Design Criteria or Specifications or in any other Contract document or is implied or inferable but no item for that work is measured or described in the Bills of Quantity and/or in the Supplementary Bill of Quantity then such work is nonetheless considered to be part of Work to be executed under this Contract.

Contractor has no right to request additional payment for such items.

Contractor shall, in performing the Work, provide all equipment, facilities, utilities, materials, transportation, supervision, design (as specified in the Scope of Work) and engineering services, Labour and all other things necessary for the efficient, safe, proper and timely execution of Work except for those supplies, works and services which shall be provided by Company as set out expressly elsewhere in this Contract.

Contractor shall, in accordance with and subject to the terms and conditions of this Contract, perform quality management and inspection activities in procurement, fabrication, construction, testing, pre-commissioning and commissioning.

At his cost and responsibility, the Contractor to acquire and secure all permits, approvals and/or licenses from all local, state or national government authorities or public service undertakings necessary for the performance of the Contract, including, but not limited into the licenses for materials and equipment to be incorporated into the Work, visas, work permits and residence permits for Contractor's Personnel, and entry permits for all Contractor's Equipment to be imported.

Provide proper security at the Work Site and at Contractor's materials yard acceptable to Company.

Prepare and submit to Company for its approval, with such promptness as not to cause any delay in the progress of Work Schedule, all the Documents including (where applicable) particular calculations, detailed design and shop drawings, necessary for the execution of the Work which shall comply with the specifications and other provisions of the Contract. Contractor shall be responsible for any discrepancies, errors or omissions in the Documents produced by him, whether such document have been approved by Company or not. Company's approval for any documents shall not release Contractor from his obligations under this Contract. Contractor shall not depart from any approved Documents without Company's approval in writing.

Provide 'as built' drawings and other project records, all in the English language, to facilitate operation and maintenance of the Work.

Participate in regular Work progress meetings to be scheduled by the Company Representative.

Perform all other obligations, work and services and furnish all other things which are required by the terms of this Contract or which can reasonably be inferred from the terms of this Contract as being necessary for the successful and timely completion of the Work.

Contractor shall observe and comply with all applicable and relevant laws, regulations and local customs in the country, where any portion of the Work is carried out whether national, provincial municipal or otherwise concerning, but not limited to customs exchange control environmental protection, safety, employment or training of local nationals, registration, import and export of equipment and materials, and taxation. Contractor shall indemnify and hold Company and its representatives and employees harmless against and from any fine and penalty of whatsoever nature resulting from Contractor's failure to so comply.

The Contractor shall carry out and be fully responsible of the Detailed Design of the parts of the Work designated in his Design Scope (Appendix to the Scope of Work), to meet as a minimum the Concept Drawings and the Design requirements, laws of the country, and other provisions of the Contract as well as sound engineering principles and good engineering practices.

The Company has the right to station, at the Company's cost, engineer(s) at the places where the detailed design and engineering work is executed to inspect the Contractor's performance of the Work and to give the Contractor explanation or clarification in regard to the specifications and requirements stipulated in the Contract. Any such explanation or clarification shall not relieve the Contractor from his complete responsibility for design and engineering in accordance with the requirements of this contract.

The Contractor shall complete the design part and engineering work under his scope and submit to the Company for its consent all the Documents including calculations, drawings, specifications, method statements, technical information and any other things required to be submitted under this Contract as a result of this design and engineering work, in accordance with the manner specified in Schedule B and within the period specified in Schedule B and/or in the approved work schedule.

Within the period specified in Schedule B, the Company shall review, inspect and examine the same and shall notify the Contractor in writing of the Company's consent or otherwise, provided that he Company may withhold its decision on its consent or otherwise for any Documents until other relevant Documents are submitted when necessary.

In the event that the Company does not consent to the use of any Documents, the Company shall specify the particular respects in which such Documents that the Company does not consent to shall be forthwith modified to meet the requirements of the Contract and re-submitted to the Company.

In the event that the Company does not consent to the use of any Documents, the Company shall specify the particular respects in which such Documents fail to comply with the requirements of the Contract. Any Documents that the Company does not consent to shall be forthwith modified to meet the requirements of the Contract and re-submitted to the Company.

The Contractor shall be responsible for any discrepancies, errors or omissions in the Documents, whether such Documents have received the Company's consent or not. The Company's consent shall not be deemed to relieve the Contractor from any of his responsibilities under this Contract.

The Contractor shall not depart from any Documents that have previously received the Company's consent, without the Company's prior consent in writing.

8.15   Contract Provisions: Company Obligations

Company shall, in accordance with and subject to the terms and conditions of this Contract, and in particular in conformance with Schedule B.

Perform Company procurement responsibilities in accordance with Schedule G.

Obtain all permits, licenses, and other governmental authorisations which official procedures require to be obtained in Company's name and which are necessary for the performance of the Work.

Allow Contractor partial and progressive access, subject to Company's normal security control and safety procedures, to the portions of the Work Site as required and in line with the Milestones in Appendix 1 to the Scope of Work for the orderly performance of the Work.

Obtain any rights-of-way that are determined by Company to be required for the performance of the Work.

Appoint one or more Company Representatives.

Perform all other obligations required of it by the terms of this Contract in such time and manner as to facilitate the orderly execution of the Work.

It is fully acknowledged by the Contractor that the term 'possession of' or 'access to' the Work Site in the context of this clause refers only to temporary permission granted by the Company to the Contractor progressively for access to the various portions of the Work Site for the sole purpose of execution of the Work in line with the Start Milestone in Attachment 1 to the Scope of Work. It is further acknowledged and fully accepted by the Contractor that the Work Site is and shall remain under the full jurisdiction of the Company who, without affecting the proper Work progress by the Contractor, may exercise such controls as he deems necessary in respect of entry to and temporary occupation of the Work Site by the Contractor or any other contractor.

Contractor shall bear all costs and charges for special or temporary rights of way required by him in connection with access to the work Site. Contractor shall also provide at his own cost any additional facilities outside the Work Site required by him for the purpose of Work.

8.16   Contract Provisions: Work Commencement, Execution and Completion

Except for preparatory work directly related to the actual physical installation of the Work. Contractor shall commence the Work promptly upon the effective date of this Contract. Contractor shall commence the remaining Work promptly upon the receipt of a written 'Notice to Proceed' from Company Representative, which notice shall, provided Contractor has complied with all other contractual obligations, be issued within 28 days of the effective date of the Contract and will state the Work Commencement Date, Critical Milestones and Completion Date of Work. Contractor shall execute Work completion with diligence and dispatch so that the Critical Milestone Dates and Time for Completion are met and exception Items are promptly completed or corrected.

The 'Notice to Proceed' will only be issued following receipt and approval of the Performance Bond pursuant to Paragraph 2.8.2 of the Schedule C and Insurance Certificate pursuant to Paragraph 16.3 of Schedule 16.3 of Schedule A.

'Practical Completion' shall be achieved for the Work, or any separable portion thereof, when the Work, or portion thereof, is constructed in strict compliance with all requirements of this Contract, including the successful completion of all required inspection, testing, and commissioning; All utility and electric power systems are fully operational; The Work Site is in a clean and safe condition, with construction debris, equipment, and excess materials removed; and Company has issued a Practical Completion Acceptance Notice (MCAN) for the Work.

Not less than 30 days prior to the date Contractor anticipates Practical Completion will be achieved, Contractor shall notify Company Representative of the anticipated date. When Contractor considers that Practical Completion has been achieved, he shall issue a written notice to that effect. Company will inspect the Work as soon as practicable. If the Work is not in accordance with this Contract, Company shall so notify Contractor in writing, specifying the respects in which the Work is deficient, and Contractor shall promptly remedy the deficiency at his expense. Following action by Contractor to remedy deficiencies, further inspection will be conducted on the same terms and conditions.

Company may, at its option, issue a 'Practical Completion Certificate' or partial 'Practical Completion Certificate' with a list of items (Exception Items) which are incomplete, defective or otherwise not in accordance with this Contract and which to not affect safe and orderly occupation and operation of the Work by Company. The issuance of such a notice shall not relieve Contractor of its obligation to complete or correct such items at no cost to Company, or preclude Company from adding additional Exception Items. When the Exception Items have been carried out and completed to the satisfaction of Company, Company will issue to Contractor a Certificate of Completion of Exception Items related to the 'Practical Completion Certificate' or partial 'Practical Completion Certificate'.

Company shall have the right to take possession of, and use for any purpose, any part of the Work at any time prior to Practical Completion after so notifying Contractor. Such taking possession or use shall not be deemed to be Company's acknowledgement of Practical Completion and shall in no way limit or waive Contractor's costs or the time required for completing the Work, Company will initiate a Change Order making any required adjustment to the Critical Milestone Dates, the Time for Completion, or the compensation due Contractor. However, should such taking possession or use result from Contractor's failure to execute the Work according to the Work Schedule, Contractor shall not be entitled to any adjustment.

The Critical Milestones and/or Time for Completion of Work will be extended if Contractor is delayed or impeded in the performance of the Work by causes expressly set out in this Contract that permit an adjustment of the Critical Milestone Dates and/or Time for Completion of Work, or by a breach of Contract or act of prevention by Company, by such reasonable period as fairly reflects the delay or impediment sustained by Contractor.

If Contractor intends to apply for an extension of a Critical Milestone and/or Time for Completion, Contractor shall give notice to Company of such intention as soon as possible and in any event within 28 days of the start of the event giving to such delay, together with any other notice required by the Contract and relevant to such event. Within 28 days of the date of issue of such notice, Contractor shall submit full supporting details of his application with an analysis of the delay impact on his Work Schedule as well as any particulars and explanations that may be requested by Company. Contractor shall make his best efforts to mitigate delays that arise. Company will reject any application for an Extension of Time for Completion that does not comply with the requirements of this Paragraph.

In the event that Contractor fails to complete the Work or a portion thereof within the required Time for Completion as may be extended pursuant to the applicable provisions in the Contract, Company shall be entitled to deduct Liquidated Damages from payments due Contractor. Such Liquidated Damages shall be computed for each day of delay after the last day of the Time for Completion of Work up to and including the date of Practical Completion as stated in Company's Practical Completion Acceptance Notice of the whole of the Work or portion thereof. The amounts and limit of Liquidated Damages are stated in the Form Agreement and shall be calculated as a percentage of the Contract Price for the whole Work.

The Liquidated Damages shall become applicable immediately upon expiry of the Time for Completion of Work without Company's serving notice to Contractor, having resource to court action, or proving that Company has actually suffered any damage or loss on account of such delay in completion of the Work or portion thereof.

If before the Time for Completion of the whole of the Work or, if applicable, any portion thereof, a Practical Completion Acceptance Notice Certificate has been issued for any part of Work or for a portion thereof, the Liquidated Damages for delay in completion of the remainder of Work or of portion thereof shall, for any period of delay after the date stated in such Practical Completion Acceptance Notice and in the absence of alternative provisions in the Contract, be reduced in the proportion which the value of the part so certified bears to the value of the whole Work or portion thereof, as applicable. The provisions of this Paragraph shall only apply to the rate of Liquidated Damages and shall not affect the limit thereof.

Company shall have the right to affect such Liquidated Damages through any means Company thinks appropriate, such as deduction from payment of any invoice and any other monies which are otherwise payable to Contractor or credited to Company as Contractor's performance assurance.

In addition to the damages, Contractor shall be obliged to bear the additional cost for the Engineer and Company Representative during the period of delays. This additional cost for the Engineer and Company Representative shall be calculated on the basis of Engineer's and Company Representative's remuneration from the Company for their services on the Contract.

8.17 Contract Provisions: Changes

At any time, Company may direct Contractor to make a change within the general scope of this Contract ('Change') such as, but not limited to additions, omissions and alterations of the Work and Contractor shall perform the Work as changed. Such changes shall be set forth as ('Change Order'). Each Change Order shall be signed by Company's Contract signatory or their authorised delegate and the authorised signatory of Contractor. All Work involved in a Change shall be performed in accordance with the terms and conditions of this Contract, and shall not otherwise affect the existing rights or obligations of the parties hereto except as expressly provided in this Contract or in a signed Change Order.

In addition to describing the Change, a Change Order shall include:

1. Any adjustment in the Critical Milestone Dates or Time for Completion resulting from the Change; and
2. The lump sum price of or the basis for determining any increase in the compensation due Contractor or credit due Company as a result of the Change, if any.

If a Change may result in a request for an adjustment in the compensation due to Contractor or a request for an adjustment to the Time for Completion of Critical Milestone Dates, Contractor shall promptly notify Company Representative orally, followed by prompt written notification. In no event shall Contractor proceed with the Work involved in the Change without a Change Order signed by Company. If Contractor proceeds with the additional Work involved in such a Change Order, Contractor shall not be entitled to any additional compensation for the Work performed or to any adjustment of the Time for Completion and/or Critical Milestone Dates as a result of the Change.

Compensation for Work performed by Contractor under a Change Order or Credit to Company for Work deleted by a Change Order shall be calculated and paid or offset in accordance with Schedule C.

Should Company and Contractor fail to agree as to the amount or method of determining adjustments in compensation due Contractor, adjustments in the Time for Completion or Critical Milestone Dates, or whether a direction from Company constitutes a Change, Company may direct Contractor, in writing, to proceed with the Work as changed and Contractor shall proceed with the Work as changed. Company shall compensate Contractor or calculate the credit due to Company in accordance with its good faith estimate of the costs or savings resulting from the Change. Contractor's performance of the Work as changed shall not prejudice its position that such direction constitutes a Change, that the Time for Completion and/or Critical Milestone Dates should be adjusted, or that Contractor should receive additional compensation for such Work; or Company's position that it is entitled to a credit. Such disputes shall be resolved in accordance with Paragraph 19 of Schedule A.

Contractor shall review and evaluate Company provided Design, the Work specifications, and the other drawings, specifications and standards referenced in this Contract to identify possible cost reductions. In the event Contractor identifies and documents a proposal ('Proposal') which will result in a net reduction in Company's total cost for the design, procurement and construction of the Work ('Net Cost Reduction') and which requires a Change Order to implement, Company may, in its sole discretion, issues a Change Order implementing the Proposal Contractor's Proposal shall contain at least a description of the existing and proposed Contract requirements and an assessment of the consequences of implementing the Proposal, including an estimate of the Net Cost Reduction.

Contractor shall provide Company, together with a Change Order Price Proposal, a sufficiently detailed breakdown of the proposed Change Order Price along with copies of all related Manufacturer/Subcontractor quotations for the materials/equipment added or deleted in the Change. Contractor shall disclose to Company any amount of discount or rebate Contractor would receive from Manufacturer(s) Subcontractor(s), directly or indirectly, for the said materials/equipments and the cost to be ultimately incurred by Contractor, directly or indirectly after all direct and indirect discounts and/or rebates to Contractor. Contractor shall enable Company to review all related records and documents.

8.18   Contract Provisions: Claims Settlement

Contractor shall inform Company, in writing, within 30 days following in occurrence of discovery, of any item or event which Contractor knows, or reasonably should know, may result in a request for an extension of time for completing the Work or any separate portion thereof, or for additional or reduced compensation under this Contract. Company and Contractor shall endeavour to satisfactory resolve the matter. Should it not be disposed of to Contractor's satisfaction, Contractor shall forthwith deliver a written notice of claim complete with all supporting documentation in triplicate to Company at its main address.

Notice of any claim of Contractor against Company for an extension of Work or any separable portion thereof, of for additional compensation of any kind under this Contract, shall be set forth in writing by Contractor and filed with Company within 30 days after failure to agree. This notice shall contain a written analysis of all elements of the claim accompanied by itemised supporting data identifying, to the extent practicable, the effect on time for completion of Work or any separable portion thereof and the amount of additional compensation claimed by Contractor. Failure by Contractor to so file such written analysis with supporting data within said 30 days period shall be deemed conclusively to be a waiver of all of Contractor's rights to any such extension of time for completion or to additional compensation.

Provided also that where an event has a continuing effect such that it is not practicable for Contractor to submit detailed particulars within the period of 30 days referred to in Paragraph above, he shall nonetheless be entitled to an extension of time or additional or reduced compensation provided that he has submitted to Company interim particulars at intervals of not more than 30 days and final particulars within 30 days of the end of the effects resulting from the event. On receipt of such interim particulars; Company shall, without undue delay, make an interim determination of extension of time or additional or reduced compensation and, on receipt of the final particulars; Company shall review all the circumstances and, as appropriate, determine on overall extension of time or additional or reduced compensation in regard to the event.

Within a reasonable time of receipt by Company of the Contractor's written analysis of the claim with supporting data, Company shall issue to Contractor a written response giving Company's determination detailing terms of settlement or rejection of the claim.

Should Contractor and Company be unable to agree upon a settlement of any claim, the matter shall be treated as an unresolved dispute in accordance with Schedule E.

Notwithstanding any other provisions of the Contract, if Contractor fails to comply with any of the provisions of the Contract, including, but not limited to the related provisions of this Schedule in respect of any event, act or omission of whatsoever nature which, in the opinion of Contractor fairly entities him to an extension of the Time for Completion and/or additional payment, then such failure shall constitute on the part of Contractor a definitive and irrevocable waiver of any entitlement and Contractor shall be effectively barred from raising any claims arising from such event, act or omission thereafter.

Should any dispute arise between Company and Contractor during Contractor's performance of Work, Contractor shall, unless Company directs otherwise, continue to perform Work and any additional Work which Company may direct Contractor to perform.

8.19 Contract Provisions: Contract Price and Payment Provisions

A full and complete compensation for Contractor's performance of the Work and all of Contractor's obligations hereunder in accordance with the terms and conditions of the Contract, Company shall pay Contractor a lump sum Contract Price as specified in the Form of Agreement as adjusted from time to time by Change Orders.

Except as otherwise provided herein below, the Contract Price constitutes the entire compensation due Contractor for the Work and all of Contractor's obligations hereunder, delivery and completion of Work at the site (including design wherever required under the Scope of Work) and includes, but is not limited to, compensation for any Government-caused cost increases imposed at any time, all applicable taxes, duties, Customs Duties, fees, overheads, profit, mobilisation and demobilisation, and all other direct and indirect costs and expenses incurred or to be incurred by Contractor hereunder.

The Contract rates in Bill of Quantity and Contract Price and rates in Supplementary Bill of Quantity are firm for the duration of the Contract and are not subject to escalation for any reason. No adjustments in the Contract rates in Bill of Quantity and Contract Price and rates in Supplementary Bill of shall be made as a result of changes in the relative value of any currency.

8.20   Calculation Methodology for Project A

The claim amounts are the actual damage that the Contractor has incurred and these amounts are obtained by subtracting the actual amount that the Contractor has incurred from the contract price which is the amount that the Contractor has priced in the original bill of quantity. This calculation methodology has been done in each Division for all the structures for the Project A.

The calculations as described were done as follows:

The Contract Cost of the Bill of Quantity per Division $= X$.

The Percentage for each Damage Heading per Division per Structure $= \%P$. Damage headings relate to: labour, equipment, subcontractors, site indirect costs or head office overheads and direct costs. This is calculated as such to show the reader how the bill of quantity can be broken up into parts and then the damages can be calculated accordingly when required. Therefore, the claimant in this global claim could have claimed for parts of the projects only where affected. Then, the damages are calculated where due to specific disruption. For instance, if delays and disruptions affected Division 3[23] and Division 15[24] for a specific structure in the Project, then the damages to these types of works could be calculated applying the modified total approach.

The Total Amount for Each Damage Headings per Division $= Y_1 = (\%P \times X)$.

The Total Cost Incurred by the Contractor for each Heading (Labour, Equipment, Sub-Contractors, Site Indirect Costs, Head Office Overhead) $= Y_2$.[25]

The Claim Amount for each Heading per Division per Structure $= Y = Y_2 - Y_1$.

The Total Claimable amount for the Project A project $= \sum Y$.

8.21   Bill of Quantity Components

The bill of quantity is divided into 16 divisions with each division having many components. The divisions and components are structured as follows:

---

[23] Division 3 relates to Concrete works.

[24] Division 15 relates to Mechanical Installation works.

[25] This is calculated from the contractor audited account or cost control department.

8.21.1 Division 1: General Requirements

Summary of work for Division 1

1.1. Cutting and patching
1.2. Reference standards
1.3. Submittals
1.4. Quality requirements
1.5. Inspection and testing services
1.6. Construction facilities and temporary controls
1.7. Security and safety
1.8. Access roads, parking areas
1.9. Traffic regulations
1.10. Field offices and sheds
1.11. Substitutions
1.12. Starting of systems
1.13. Contract closeout
1.14. Cleaning
1.15. Operation and maintenance data
1.16. Selective demolition

8.21.2 Division 2: Site Work

Summary of Work for Division 2

2.1. Soil investigation
2.2. Cavity probing
2.3. Demolition
2.4. Interceptors
2.5. Clearing and grubbing
2.6. Earthworks
2.7. Compaction and testing of earthwork
2.8. Structural excavation and back fill
2.9. Trenching, back filling, compaction and general grading
2.10. Aggregate or granular sub-base
2.11. Aggregate base course
2.12. Filter fabric
2.13. Termite control
2.14. Slope protection
2.15. Concrete piles
2.16. Domestic water distribution
2.17. Pavements and asphaltic concrete
2.18. Slurry seal
2.19. Pavers

2.20. Precast concrete curbs
2.21. Septic tank
2.22. Pavements markings
2.23. Traffic signage
2.24. Site work expansion joints
2.25. Storm water drainage
2.26. Manhole covers and frames
2.27. Sanitary drainage
2.28. Pumping stations and pumping mains
2.29. Sewage treatment plant
2.30. Underground piping
2.31. Underslab drainage
2.32. Subsoil drainage
2.33. Playground surfacing
2.34. Irrigation
2.35. Exterior pools and fountains
2.36. Chain link fence and gates
2.37. Traffic signals
2.38. Play field equipment and structures
2.39. Trees, shrubs, ground cover and grass

8.21.3 Division 3: Concrete

Summary of Work for Division 3

3.1. Formwork
3.2. Concrete reinforcement
3.3. Concrete accessories
3.4. Cast in place concrete
3.5. Exposed aggregate concrete finish
3.6. Stamped concrete or imprinted concrete finish
3.7. Concrete curing
3.8. Precast concrete
3.9. Precast concrete hollow core planks
3.10. Architectural precast concrete
3.11. Reinforced aerated concrete panels
3.12. Glass fiber reinforced concrete
3.13. Lightweight concrete fill
3.14. Concrete floor topping
3.15. Cement-based screed
3.16. Concrete repair

8.21.4 Division 4: Masonry

Summary of Work for Division 4

4.1. Mortar and masonry grout
4.2. Glass unit masonry
4.3. Unit masonry

8.21.5 Division 5: Metals

Summary of Work for Division 5

5.1. Metal fastenings
5.2. Structural steel
5.3. Space framing
5.4. Metal decking
5.5. Miscellaneous metal
5.6. Metal stairs
5.7. Handrails and railings
5.8. Gratings and floor plates
5.9. Ornamental metal
5.10. Ornamental handrails and railings
5.11. Expansion joint assemblies

8.21.6 Division 6: Wood and Plastic

Summary of Work for Division 6

6.1. Rough carpentry
6.2. Finish carpentry
6.3. Custom casework
6.4. Panel work

8.21.7 Division 7: Thermal and Moisture Protection

Summary of Work for Division 7

7.1. Bituminous membrane waterproofing
7.2. Sheet membrane waterproofing
7.3. Bituminous, damp proofing and waterproofing
7.4. Board insulation

7.5. Sprayed fire proofing
7.6. Fire stopping
7.7. Metal panels
7.8. Membrane roofing
7.9. Modified protected membrane
7.10. Sheet metal roofing
7.11. Sheet metal flashing and trim
7.12. Roof hatches
7.13. Joint sealers

8.21.8 Division 8: Doors and Windows

Summary of Work for Division 8

8.1. Steel doors
8.2. Steel frames
8.3. Aluminium doors and frames
8.4. Sliding alumium framed glass doors
8.5. Wood doors and frames
8.6. Flush wood doors
8.7. Access doors and frames
8.8. Steel detention doors and frames
8.9. Overhead cooling roller shutter doors
8.10. Interior glass wall system
8.11. All glass entrance
8.12. Aluminum windows
8.13. Security windows
8.14. Metal framed skylights
8.15. Door hardware
8.16. Automatic door equipment
8.17. Detention door hardware
8.18. Glass and glazing
8.19. Mirrors
8.20. Glazed aluminum curtain wall system

8.21.9 Division 9: Finishing

Summary of Work for Division 9

9.1. Metal stud framing, furring and lathing
9.2. Gypsum plaster
9.3. Veneer plaster system
9.4. Portland cement plaster

9.5. Gypsum and cement board system
9.6. Ceramic tiles
9.7. Terrazzo
9.8. Acoustic ceilings
9.9. Linear metal ceiling
9.10. Wood flooring
9.11. Stone flooring
9.12. Resilient flooring
9.13. Carpet
9.14. Interior stone facing
9.15. Wall covering

8.21.10 Division 10: Specialties

Summary of Work for Division 10

10.1. Visual display boards
10.2. Telephone specialties and projection screen
10.3. Toilet compartments
10.4. Metal louvers
10.5. Architectural screen
10.6. Wall and corner guard
10.7. Access flooring
10.8. Signage
10.9. Metal lockers
10.10. Fire protection specialties
10.11. Wire mesh partitions
10.12. Site-assembled demountable partitions
10.13. Operable panel partitions
10.14. Metal storage shelving
10.15. Toilet and bath accessories

8.21.11 Division 11: Equipment

Summary of Work for Division 11

11.1. Maintenance equipment
11.2. Vault door
11.3. Library equipment
11.4. Theatre and stage equipment
11.5. Traffic control equipment
11.6. Loading dock equipment
11.7. Waste compactors

11.8. Food service and laundry equipment
11.9. Hood and ventilation equipment
11.10. Exercise equipment
11.11. Laboratory fume hoods
11.12. Laboratory equipment
11.13. Workshop equipment

8.21.12 Division 12: Furnishing

Summary of Work for Division 12

12.1. Metal casework
12.2. Laboratory casework
12.3. Residential casework
12.4. Drapery and tracks
12.5. Roller shades for windows
12.6. Horizontal and vertical louver blinds
12.7. Auditorium and theatre seating
12.8. Floor mats and frames
12.9. Interior plants

8.21.13 Division 13: Special Construction

Summary of Work for Division 13

13.1. Outdoor sports courts
13.2. Indoor sports
13.3. Cold store rooms
13.4. Saunas
13.5. Radiation protection
13.6. Pre-engineered steel buildings
13.7. Cable-supported structures
13.8. Hot tubes and whirlpool
13.9. Floating dock
13.10. Measurement and control instrumentation

8.21.14 Division 14: Conveying System

Summary of Work for Division 14

14.1. Electric traction elevators
14.2. Overhead traveling cranes

8.21.15   Division 15: Mechanical Installations

Summary of Work for Division 15

15.1. Basic mechanical requirements
15.2. Mechanical sound, vibration and seismic control
15.3. Electrical requirements for mechanical equipment
15.4. Building management system and automatic controls
15.5. Mechanical identification
15.6. HVAC thermal insulation
15.7. Fire protection piping
15.8. Fire protection valves
15.9. Fire protection supports, hangers and brackets
15.10. Fire pumps
15.11. Fire protection system and equipment
15.12. Pre-engineered wet chemical extinguishing systems
15.13. Plumbing piping
15.14. Plumbing valves
15.15. Plumbing supports, hangers and brackets
15.16. Plumbing specialties
15.17. Plumbing fixtures
15.18. Water recycling plant
15.19. Water heater
15.20. Tanks
15.21. Plumbing pumps
15.22. Pool and fountain equipment
15.23. Laboratory air and vacuum piping
15.24. Laboratory air and vacuum equipment
15.25. General service compressed air equipment
15.26. Diesel oil storage and piping systems
15.27. LPG storage and piping systems
15.28. Compressed air and piping systems
15.29. Welding gas and compressed air piping
15.30. Fuel gas piping
15.31. Distilled water system
15.32. Chemical waste piping
15.33. Heating, ventilating and air conditioning
15.34. HVAC noise control
15.35. Chemical water treatment
15.36. Refrigeration equipment
15.37. Refrigeration compressors
15.38. Refrigeration condensing unit
15.39. Water chillers
15.40. Close control air-conditioning units
15.41. Chilled water fan coil units; with electric heater

15.42. Air handling units
15.43. Air movers: centrifugal and axial air cleaning
15.44. Air cleaning devices
15.45. Ductwork
15.46. Ductwork accessories
15.47. Air terminal variable volume boxes
15.48. Air outlets and inlets
15.49. HVAC testing, adjusting and balancing

8.21.16 Division 16: Electrical Installations

Summary of Work for Division 16

16.1. Basic electrical requirements
16.2. Electrical boxes and fittings
16.3. Equipment connections and supports
16.4. Underground electrical services
16.5. Wiring devices
16.6. Dimming devices
16.7. Disconnect switches
16.8. Electrical metres
16.9. Power transformers
16.10. Dry type transformer
16.11. Package type sub-station
16.12. Power factor capacitors
16.13. Panel boards
16.14. Interior lighting
16.15. Exterior and street lighting
16.16. Emergency and exit lighting
16.17. Enclosed transfer switches
16.18. Un-interruptible power supply system
16.19. Earthing system
16.20. Lighting protection system
16.21. Telephone systems equipment
16.22. Voice and data cabling
16.23. Computer network
16.24. Fire alarm and detection system
16.25. Security system
16.26. Public address sound system
16.27. Master clock system
16.28. Television cabling system
16.29. Video head end system
16.30. Communication grounding system

16.31. Audio visual system
16.32. Electrical testing and commissioning.

8.22 Claims Components and Calculation

The following are the claim amounts per structure for Project A. As mentioned
earlier in this chapter, this claim is based on the assumption that the claimant is
basing his calculation on the whole of the project. This breakdown of losses for
each structure, per division and per claim heading is to demonstrate how the
claimant can potentially apply the modified total cost method to claim for specific
items in a particular project.

The bill of quantity is structured as follows:

Phase 1: Project A Building: B100

8.22.1 Division 2/Site Works

Labour cost to be claimed = 237,158 USD. Labour cost constitutes 15% of the
total cost of Division 2 as estimated by the Contractor when pricing this project.

Equipment cost to be claimed = 632,421.25 USD. Equipment cost constitutes
40% of the total cost of Division 2 as estimated by the Contractor when pricing
this project.

Subcontractors cost to be claimed = 473,316 USD. Subcontractors cost constitutes 30% of the total cost of Division 2 as estimated by the Contractor when
pricing this project.

Indirect cost to be claimed = 189,726.5 USD. Indirect cost constitutes 12% of
the total cost of Division 2 as estimated by the Contractor when pricing this
project.

Direct cost to be claimed = 47,431.5 USD. Direct cost constitutes 3% of the
total cost of Division 2 as estimated by the Contractor when pricing this project.

Total claimable amount for Division 2/Project A Building B100 =
1,581,053.25 USD.

8.22.2 Division 3/Concrete

Labour cost to be claimed = 1,062,668.75 USD. Labour cost constitutes 28% of
the total cost of Division 3 as estimated by the Contractor when pricing this
project.

Equipment cost to be claimed = 834,954 USD. Equipment cost constitutes
22% of the total cost of Division 3 as estimated by the Contractor when pricing
this project.

Subcontractors cost to be claimed = 1,328,336 USD. Subcontractors cost constitutes 35% of the total cost of Division 3 as estimated by the Contractor when pricing this project.

Indirect cost to be claimed = 455,429.5 USD. Indirect cost constitutes 12% of the total cost of Division 3 as estimated by the Contractor when pricing this project.

Direct cost to be claimed = 113,857.25 USD. Direct cost constitutes 3% of the total cost of Division 3 as estimated by the Contractor when pricing this project.

Total claimable amount for Division 3/Project A Building B100 = 3,795,245.5 USD.

8.22.3 Division 4/Masonry

Labour cost to be claimed = 41,136 USD. Labour cost constitutes 33% of the total cost of Division 4 as estimated by the Contractor when pricing this project.

Equipment cost to be claimed = 18,698.25 USD. Equipment cost constitutes 15% of the total cost of Division 4 as estimated by the Contractor when pricing this project.

Subcontractors cost to be claimed = 46,122 USD. Subcontractors cost constitutes 37% of the total cost of Division 4 as estimated by the Contractor when pricing this project.

Indirect cost to be claimed = 14,958.5 USD. Indirect cost constitutes 12% of the total cost of Division 4 as estimated by the Contractor when pricing this project.

Direct cost to be claimed = 3,739.75 USD. Direct cost constitutes 3% of the total cost of Division 4 as estimated by the Contractor when pricing this project.

Total claimable amount for Division 4/Project A Building B100 = 124,654.25 USD.

8.22.4 Division 5/Metal

Labour cost to be claimed = 101,906 USD. Labour cost constitutes 27% of the total cost of Division 5 as estimated by the Contractor when pricing this project.

Equipment cost to be claimed = 45,291.5 USD. Equipment cost constitutes 12% of the total cost of Division 5 as estimated by the Contractor when pricing this project.

Subcontractors cost to be claimed = 173,617.5 USD. Subcontractors cost constitutes 46% of the total cost of Division 5 as estimated by the Contractor when pricing this project.

Indirect cost to be claimed = 45,291.5 USD. Indirect cost constitutes 12% of the total cost of Division 5 as estimated by the Contractor when pricing this project.

Direct cost to be claimed = 11,323 USD. Indirect cost constitutes 3% of the total cost of Division 5 as estimated by the Contractor when pricing this project.

Total claimable amount for Division 5 Project A/Building A B100 = 377,429.5 USD.

8.22.5 Division 6/Wood & Plastic

Labour cost to be claimed = 34,328.25 USD. Labour cost constitutes 33% of the total cost of Division 6 as estimated by the Contractor when pricing this project.

Equipment cost to be claimed = 12,483 USD. Equipment cost constitutes 12% of the total cost of Division 6 as estimated by the Contractor when pricing this project.

Subcontractors cost to be claimed = 41,610 USD. Subcontractors cost constitutes 40% of the total cost of Division 6 as estimated by the Contractor when pricing this project.

Indirect cost to be claimed = 12,483 USD. Indirect cost constitutes 12% of the total cost of Division 6 as estimated by the Contractor when pricing this project.

Direct cost to be claimed = 3,120.75 USD. Direct cost constitutes 3% of the total cost of Division 6 as estimated by the Contractor when pricing this project.

Total claimable amount for Division 6/Project A Building B100 = 104,025 USD.

8.22.6 Division 7/Thermal & Moisture Protection

Labour cost to be claimed = 482,463.25 USD. Labour cost constitutes 35% of the total cost of Division 7 as estimated by the Contractor when pricing this project.

Equipment cost to be claimed = 165,416 USD. Equipment cost constitutes 12% of the total cost of Division 7 as estimated by the Contractor when pricing this project.

Subcontractors cost to be claimed = 523,817.25 USD. Subcontractors cost constitutes 38% of the total cost of Division 7 as estimated by the Contractor when pricing this project.

Indirect cost to be claimed = 165,416 USD. Indirect cost constitutes 12% of the total cost of Division 7 as estimated by the Contractor when pricing this project.

Direct cost to be claimed = 41,354 USD. Direct cost constitutes 3% of the total cost of Division 7 as estimated by the Contractor when pricing this project.

Total claimable amount for Division 7/Project A Building B100 = 1,378,466.25 USD.

8.22.7 Division 8/Door & Windows

Labour cost to be claimed = 1,133,115 USD. Labour cost constitutes 33% of the total cost of Division 8 as estimated by the Contractor when pricing this project.

Equipment cost to be claimed = 412,041.75 USD. Equipment cost constitutes 12% of the total cost of Division 8 as estimated by the Contractor when pricing this project.

Subcontractors cost to be claimed = 1,373,472.75 USD. Subcontractors cost constitutes 40% of the total cost of Division 8 as estimated by the Contractor when pricing this project.

Indirect cost to be claimed = 412,041.75 USD. Indirect cost constitutes 12% of the total cost of Division 8 as estimated by the Contractor when pricing this project.

Direct cost to be claimed = 103,010.5 USD. Direct cost constitutes 3% of the total cost of Division 8 as estimated by the Contractor when pricing this project.

Total claimable amount for Division 8/Project A Building B100 = 3,433,681.75 USD.

8.22.8 Division 9/Finishes

Labour cost to be claimed = 909,798.5 USD. Labour cost constitutes 43% of the total cost of Division 9 as estimated by the Contractor when pricing this project.

Equipment cost to be claimed = 253,897.25 USD. Equipment cost constitutes 12% of the total cost of Division 9 as estimated by the Contractor when pricing this project.

Subcontractors cost to be claimed = 634,743 USD. Subcontractors cost constitutes 30% of the total cost of Division 9 as estimated by the Contractor when pricing this project.

Indirect cost to be claimed = 253,897.25 USD. Indirect cost constitutes 12% of the total cost of Division 9 as estimated by the Contractor when pricing this project.

Direct cost to be claimed = 63,474.25 USD. Direct cost constitutes 3% of the total cost of Division 9 as estimated by the Contractor when pricing this project.

Total claimable amount for Division 9/Project A Building B100 = 2,115,810.25 USD.

8.22.9 Division 10/Specialties

Labour cost to be claimed = 27,414.25 USD. Labour cost constitutes 28% of the total cost of Division 10 as estimated by the Contractor when pricing this project.

Equipment cost to be claimed = 11,749 USD. Equipment cost constitutes 12% of the total cost of Division 10 as estimated by the Contractor when pricing this project.

Subcontractors cost to be claimed = 44,058.75 USD. Subcontractors cost constitutes 45% of the total cost of Division 10 as estimated by the Contractor when pricing this project.

Indirect cost to be claimed = 11,749 USD. Indirect cost constitutes 12% of the total cost of Division 10 as estimated by the Contractor when pricing this project.

Direct cost to be claimed = 2,937.25 USD. Direct cost constitutes 3% of the total cost of Division 10 as estimated by the Contractor when pricing this project.

Total claimable amount for Division 10 Project/A Building B100 = 97,908.25 USD.

8.22.10 Division 11/Equipment

Labour cost to be claimed = 134,453.75 USD. Labour cost constitutes 35% of the total cost of Division 11 as estimated by the Contractor when pricing this project.

Equipment cost to be claimed = 46,098.5 USD. Equipment cost constitutes 12% of the total cost of Division 11 as estimated by the Contractor when pricing this project.

Subcontractors cost to be claimed = 145,978.5 USD. Subcontractors cost constitutes 38% of the total cost of Division 11 as estimated by the Contractor when pricing this project.

Indirect cost to be claimed = 46,098.5 USD. Indirect cost constitutes 12% of the total cost of Division 11 as estimated by the Contractor when pricing this project.

Direct cost to be claimed = 11,524.5 USD. Direct cost constitutes 3% of the total cost of Division 11 as estimated by the Contractor when pricing this project.

Total claimable amount for Division 11/Project A Building B100 = 384,153.75 USD.

8.22.11 Division 12/Special Construction

Not applicable as there are no works for B100 that have items included in the bill of quantity and no works have been executed for this category.

8.22.12 Division 13/Special Construction

Labour cost to be claimed = 2,152.75 USD. Labour cost constitutes 25% of the total cost of Division 13 as estimated by the Contractor when pricing this project.

Equipment cost to be claimed = 1,033.25 USD. Equipment cost constitutes 12% of the total cost of Division 13 as estimated by the Contractor when pricing this project.

Subcontractors cost to be claimed = 4,133.25 USD. Subcontractors cost constitutes 48% of the total cost of Division 13 as estimated by the Contractor when pricing this project.

Indirect cost to be claimed = 1,033.25 USD. Indirect cost constitutes 12% of the total cost of Division 13 as estimated by the Contractor when pricing this project.

Direct cost to be claimed = 258.25 USD. Direct cost constitutes 3% of the total cost of Division 13 as estimated by the Contractor when pricing this project.

Total claimable amount for Division 13/Project A Building B100 = 8,610.75 USD.

8.22.13 Division 14/Conveying System

Labour cost to be claimed = 55,997.5 USD. Labour cost constitutes 22% of the total cost of Division 14 as estimated by the Contractor when pricing this project.

Equipment cost to be claimed = 12,726.75 USD. Equipment cost constitutes 5% of the total cost of Division 14 as estimated by the Contractor when pricing this project.

Subcontractors cost to be claimed = 147,629.5 USD. Subcontractors cost constitutes 58% of the total cost of Division 14 as estimated by the Contractor when pricing this project.

Indirect cost to be claimed = 30,544 USD. Indirect cost constitutes 12% of the total cost of Division 14 as estimated by the Contractor when pricing this project.

Direct cost to be claimed = 7,636 USD. Direct cost constitutes 3% of the total cost of Division 14 as estimated by the Contractor when pricing this project.

Total claimable amount for Division 14/Project A Building B100 = 254,533.75 USD.

8.22.14 Division 15/Mechanical Installations

Labour cost to be claimed = 946,088.25 USD. Labour cost constitutes 33% of the total cost of Division 15 as estimated by the Contractor when pricing this project.

Equipment cost to be claimed = 344,032 USD. Equipment cost constitutes 12% of the total cost of Division 15 as estimated by the Contractor when pricing this project.

Subcontractors cost to be claimed = 1,146,773.75 USD. Subcontractors cost constitutes 40% of the total cost of Division 15 as estimated by the Contractor when pricing this project.

Indirect cost to be claimed = 344,032 USD. Indirect cost constitutes 12% of the total cost of Division 15 as estimated by the Contractor when pricing this project.

Direct cost to be claimed = 86,008 USD. Direct cost constitutes 3% of the total cost of Division 15 as estimated by the Contractor when pricing this project.

Total claimable amount for Division 15/Project A Building B100 = 2,866,934 USD.

8.22.15 Division 16/Mechanical Installations

Labour cost to be claimed = 853,818.5 USD. Labour cost constitutes 33% of the total cost of Division 16 as estimated by the Contractor when pricing this project.

Equipment cost to be claimed = 310,479.5 USD. Equipment cost constitutes 12% of the total cost of Division 16 as estimated by the Contractor when pricing this project.

Subcontractors cost to be claimed = 310,479.5 USD. Subcontractors cost constitutes 40% of the total cost of Division 16 as estimated by the Contractor when pricing this project.

Indirect cost to be claimed = 310,479.5 USD. Indirect cost constitutes 12% of the total cost of Division 16 as estimated by the Contractor when pricing this project.

Direct cost to be claimed = 77,687.25 USD. Direct cost constitutes 3% of the total cost of Division 16 as estimated by the Contractor when pricing this project.

Total claimable amount for Division 16/Project A Building B100 = 2587,328.5 USD.

Total Claimable Amount for Phase 1—Building B100 = 19,109,835 USD (Table 8.20).

8.22.16 Phase I: Site Development

Phase I
  Site Development

8.22.17 Division 2/Site Works

Labour Cost to be claimed = 129,032 USD. Labour cost constitutes 15% of the total cost of Division 2 as estimated by the Contractor when pricing this project.

Equipment Cost to be claimed = 344,085 USD. Equipment cost constitutes 40% of the total cost of Division 2 as estimated by the Contractor when pricing this project.

Subcontractor Cost to be claimed = 258,063.75 USD. Subcontractor cost constitutes 30% of the total cost of Division 2 as estimated by the Contractor when pricing this project.

Table 8.20 Breakdown of the total claimable amount for Phase 1—Building B100

	Labour	Equipment	Sub-contract	Site indirect overhead	Head office overhead	Total losses
Project A Building (100)—all amounts in USD						
Div 2	15%	40%	30%	12%	3%	
	237,158	632,421.25	474,316	189,726.5	47,431.5	1,581,053.25
Div 3	28%	22%	35%	12%	3%	
	1,062,668.75	834,954	1,328,336	455,429.5	113,857.25	3,795,254.5
Div 4	33%	15%	37%	12%	3%	
	41,136	18,698.25	46,122	14,958.5	3,739.75	124,654.25
Div 5	27%	12%	46%	12%	3%	
	101,906	45,291.5	173,617.5	45,291.5	11,323	377,429.5
Div 6	33%	12%	40%	12%	3%	
	34,328.25	12,483	41,610	12,483	3,120.75	104,025
Div 7	35%	12%	38%	12%	3%	
	482,463.25	165,416	523,817.25	165,416	41,354	1,378,466.25
Div 8	33%	12%	40%	12%	3%	
	1,133,115	412,041.75	1,373,472.75	412,041.75	103,010.5	3,433,681.75
Div 9	43%	12%	30%	12%	3%	
	909,798.5	253,897.25	634,743	253,897.25	63,474	2,115,810.25
Div 10	28%	12%	45%	12%	3%	
	27,414.25	11,749	44,058.75	11,749	2,937.25	97,908.25
Div 11	35%	12%	38%	12%	3%	
	134,453.75	46,098.5	145,978.5	46,098.5	11,524.5	384,153.75
Div 13	25%	12%	48%	12%	3%	
	2,152.75	1,033.25	4,133.25	1,033.25	258.25	8,610.75
Div 14	22%	5%	58%	12%	3%	
	55,997.5	12,726.75	147,629.5	30,544	7,636	254,533.75
Div 15	33%	12%	40%	12%	3%	
	946,088.25	344,032	1,146,773.75	344,032	86,008	2,866,934
Div 16	33%	12%	40%	12%	3%	
	853,818.5	310,479.5	1,034,931.25	310,479.5	77,619.75	2,587,328.5

Indirect Cost to be claimed = 103,225.5 USD. Indirect cost constitutes 12% of the total cost of Division 2 as estimated by the Contractor when pricing this project.

Direct Cost to be claimed = 25,806.5 USD. Direct cost constitutes 12% of the total cost of Division 2 as estimated by the Contractor when pricing this project.

Total claimable amount for Division 2 Site Development 860,212.75 USD.

8.22.18 Division 3/Concrete

Labour Cost to be claimed = 1,015.75 USD. Labour cost constitutes 28% of the total cost of Division 3 as estimated by the Contractor when pricing this project.

Equipment Cost to be claimed = 798 USD. Equipment cost constitutes 22% of the total cost of Division 3 as estimated by the Contractor when pricing this project.

Subcontractor Cost to be claimed = 1,269.75 USD. Subcontractor cost constitutes 35% of the total cost of Division 3 as estimated by the Contractor when pricing this project.

Indirect Cost to be claimed = 428.5 USD. Indirect cost constitutes 12% of the total cost of Division 3 as estimated by the Contractor when pricing this project.

Direct Cost to be claimed = 108.75 USD. Direct cost constitutes 30% of the total cost of Division 3 as estimated by the Contractor when pricing this project.

Total claimable amount for Division 3/Site Development = 3,627.75 USD.

8.22.19 Division 4/Masonry

Not applicable as there are no works for Site Development that have items included in the bill of quantity and no works have been executed for this category.

8.22.20 Division 5/Metal

Labour Cost to be claimed = 399.25 USD. Labour cost constitutes 27% of the total cost of Division 5 as estimated by the Contractor when pricing this project.

Equipment Cost to be claimed = 177.5 USD. Equipment cost constitutes 22% of the total cost of Division 5 as estimated by the Contractor when pricing this project.

Subcontractor Cost to be claimed = 680.25 USD. Subcontractor cost constitutes 46% of the total cost of Division 5 as estimated by the Contractor when pricing this project.

Indirect Cost to be claimed = 177.5 USD. Indirect cost constitutes 12% of the total cost of Division 5 as estimated by the Contractor when pricing this project.

Direct Cost to be claimed = 44.25 USD. Direct cost constitutes 12% of the total cost of Division 5 as estimated by the Contractor when pricing this project.

Total claimable amount for Division 5/Site Development 1,478.75 USD.

8.22.21 Division 6/Wood and Plastic

Not applicable as there are no works for Site Development that have items included in the bill of quantity and no works have been executed for this category.

8.22.22 Division 7/Thermal & Moisture Protection

Not applicable as there are no works for Site Development that have items included in the bill of quantity and no works have been executed for this category.

8.22.23 Division 8/Doors and Windows

Not applicable as there are no works for Site Development that have items included in the bill of quantity and no works have been executed for this category.

8.22.24 Division 9/Finishings

Not applicable as there are no works for Site Development that have items included in the bill of quantity and no works have been executed for this category.

8.22.25 Division 10/Specialties

Not applicable as there are no works for Site Development that have items included in the bill of quantity and no works have been executed for this category.

8.22.26 Division 11/Equipment

Not applicable as there are no works for Site Development that have items included in the bill of quantity and no works have been executed for this category.

8.22.27 Division 12/Furnishing

Not applicable as there are no works for Site Development that have items included in the bill of quantity and no works have been executed for this category.

8.22.28 Division 13/Special Construction

Labour Cost to be claimed = 160,105.25 USD. Labour cost constitutes 25% of the total cost of Division 13 as estimated by the Contractor when pricing this project.

Equipment Cost to be claimed = 76,850.5 USD. Equipment cost constitutes 12% of the total cost of Division 13 as estimated by the Contractor when pricing this project.

Subcontractor Cost to be claimed = 307,402 USD. Subcontractor cost constitutes 48% of the total cost of Division 13 as estimated by the Contractor when pricing this project.

Indirect Cost to be claimed = 76,850.5 USD. Indirect cost constitutes 12% of the total cost of Division 13 as estimated by the Contractor when pricing this project.

Direct Cost to be claimed = 19,212.5 USD. Direct cost constitutes 12% of the total cost of Division 13 as estimated by the Contractor when pricing this project.

Total claimable amount for Division 13/Site Development = 640,420.75 USD.

8.22.29 Division 15/Mechanical Installations

Labour Cost to be claimed = 83,463.75 USD. Labour cost constitutes 33% of the total cost of Division 15 as estimated by the Contractor when pricing this project.

Equipment Cost to be claimed = 30,350.5 USD. Labour cost constitutes 12% of the total cost of Division 15 as estimated by the Contractor when pricing this project.

Subcontractor Cost to be claimed = 101,168.25 USD. Subcontractor cost constitutes 40% of the total cost of Division 15 as estimated by the Contractor when pricing this project.

Indirect Cost to be claimed = 30,350.5 USD. Subcontractor cost constitutes 12% of the total cost of Division 15 as estimated by the Contractor when pricing this project.

Direct Cost to be claimed = 7,587.5 USD. Subcontractor cost constitutes 3% of the total cost of Division 15 as estimated by the Contractor when pricing this project.

Total claimable amount for Division 15/Site Development 252,920.5 USD.

8.22.30 Division 16/Mechanical Installations

Labour Cost to be claimed = 95,860.25 USD. Labour cost constitutes 33% of the total cost of Division 16 as estimated by the Contractor when pricing this project.

Equipment Cost to be claimed = 34,858.25 USD. Equipment cost constitutes 12% of the total cost of Division 16 as estimated by the Contractor when pricing this project.

Subcontractor Cost to be claimed = 116,194.25 USD. Subcontractor cost constitutes 40% of the total cost of Division 16 as estimated by the Contractor when pricing this project.

Table 8.21 Breakdown of the total claimable amount for Phase 1—Site development

	Labour	Equipment	Sub-contract	Site indirect overhead	Head office overhead	Total
Site Development (Project A)—all amounts in USD						
Div 2	**15%**	**40%**	**30%**	**12%**	**3%**	
	129,032	344,085	258,063.75	103,225.5	25,806.5	860,212.75
Div 3	**28%**	**22%**	**35%**	**12%**	**3%**	
	1,015.75	798	1,269.75	435.25	108.75	3,627.75
Div 5	**27%**	**12%**	**46%**	**12%**	**3%**	
	399.25	177.5	680.25	177.5	44.25	1,478.75
Div 13	**25%**	**12%**	**48%**	**12%**	**3%**	
	160,105.25	76,850.5	307,402	76,850.5	19,212.5	640,420.75
Div 15	**33%**	**12%**	**40%**	**12%**	**3%**	
	83,463.75	30,350.5	101,168.25	30,350.5	7,587.5	252,920.5
Div 16	**33%**	**12%**	**40%**	**12%**	**3%**	
	95,860.25	34,858.25	116,194.25	34,858.25	8,714.5	290,485.5

Indirect Cost to be claimed = 34,858.25 USD. Indirect cost constitutes 12% of the total cost of Division 16 as estimated by the Contractor when pricing this project.

Direct Cost to be claimed = 8,714.5 Direct cost constitutes 3% of the total cost of Division 16 as estimated by the Contractor when pricing this project.

Total claimable amount for Division 16/Site Development = 290,485.5 USD.

Total Claimable Amount for Phase 1-Site Development = 2,049,146 USD (Table 8.21).

8.23 Total Claimable Amount

By following the same calculation method for rest of Project A, the breakdown of the global claim per structure are summarised in the Table 8.1.

As mentioned earlier in this case-study chapter, the contractor has adjusted his losses by subtracting from this claim the variations agreed to be paid by the Client which amounts to 394,870 USD and the other claim that he will be submitting in relation to the materials cost increase which will total to 2,440,480 USD. Therefore this global claim amount falls under the modified total cost method and is summarised as follows:

Total losses = 45,519,643 USD
Agreed variations = 394,870 USD
Material escalation claim = 2,440,480 USD (This claim will be submitted separately)
Global claim amount = 42,684,293 USD (This is the amount submitted in this claim).

8.24 Summary

This is global claim relating to the Project A with a value of 42,684,293 USD. This claim includes the legal and contractual background that the Contractor is basing his right for compensation and which have imposed a heavy burden on the operation of the project and have put the Contractor under extreme economic duress due to the heavy losses that he has incurred and still incurring. Project A was marred with numerous problems. The number and impact of these problems are rarely encountered in one single project and if only few of these problems occur in one particular project can inflict heavy losses on the contractor.

Each claim heading is written in detail and all the legal background and the problems encountered on the Project are supported with facts and documents.[26] The Contractor seeking remedies and relief of the problems encountered by applying certain doctrines in law such as misrepresentation, economic duress and frustration which allow a degree of release of the terms of the contract or adjustment of these terms in the face of the exceptional unforeseen conditions that occurred through the life of the Project.

The delays and obstacles that occurred could not be foreseeable by the Contractor or any other experienced contractor and where outside the scope of the assumed risks considered under the Contract at the time the job was awarded in October 2006 and during the period prior to that date when the job was estimated and the plans to construct were put in place. The consequence of the above disruptions is that the contract could be rescinded due to the doctrine of frustration where, in the event of frustration, the contract will be discharged in relation to future performance obligations of both parties. The Contractor, even knowing his right to rescind the contract due to these circumstances decided to continue the work even though under heavy losses amounting to more than 40 million USD at present and the Project being delayed beyond any reasonable date that could have been envisaged and foreseeable. The Contractor faced enormous difficulties from the outset and these are explained in great detail in the claim document. A summary of these difficulties falls under the following headings:

8.24.1 Late Arrival of the Consultants Team to Site

In Project A, the arrival of the consultants site-based teams, in the numbers required to facilitate the required early momentum, was very slow and this had a dramatic negative effect on the Contractor planned rate of output in the early months. It is also clear from the body of this claim document that the consultant team failed to provide the early assistance required.

---

[26] Not included in this book for the sheer number.

8.24.2 Effect of the Construction Boom

This phenomenon had two dramatically negative effects on the Project. Firstly, the tender allowances for resources were fully consumed far earlier than planned and secondly, the difficulty in securing sufficient resources to satisfy the requirements of the planned activities severely degraded the Contractor's planned output and progress of work.

8.24.3 Effect of the Remoteness of the Site

The factor of the remoteness of the site is closely linked to the construction boom because fierce demand for labour resources meant that the labour pool could be more selective in terms of choosing more palatable locations. Consequently, the difficulty in securing sufficient resources to satisfy the requirements of the planned activities severely degraded the Contractor's planned output and progress of work.

8.24.4 Effect of Dealing with the Client Structure, the Site Safety and the Security Regime

The contract documents were, at best, ambiguous in this regard. In reality, there is a measure of misrepresentation as to the *modus operandi* of the site safety and security regime procedures. Layers of bureaucratic red tape had to be navigated to affect simple tasks relating to receiving instructions, entering the site, obtaining passes for labour and subcontractors, site safety and site security. In addition to the direct difficulties associated with dealing with the Client structure, the disruptive effect of the site safety and the security regime procedures on-site moral, productivity and effectiveness were significant. The requirement for the Contractor to conform and to accept the Client imposed changes were invariably dictated by applying coercive pressures through these demanding procedures.

8.24.5 Disruption Arising from Change Orders and Disputed Change Orders

A considerable volume of change orders were issued during the construction period and these changes, combined with the very considerable number of disputed change orders, as detailed in the body of this claim document, caused considerable cumulative disruption to the Contractor's regular progress of work. The Contractor asserts that the number, timing and effect of the changes issued impacted his ability to plan and perform the work. The Client caused disruptions beyond the Contractor control and foreseeability. Multiple change orders have been issued on this Project; and there is a solid *prima facie* case showing that the Client's numerous change orders form a basis for the Contractor to claim as a result of their cumulative and disruptive impact.

8.24.6 Effect of Delayed Approvals [Drawings and Submittals]

The actual history of the consultant response time on the Project was very poor and was an important contributory factor to the disruption and delay experienced by the Contractor. The Contractor makes this assessment on the basis of analysing the return of submittals within three categories which are (1) Greater than 15 days but less than 25 days; (2) Greater than 25 days but less than 50 days; and (3) Greater than 50 days. With respect to the first category 7,842 drawing submittals took more than 15 days for the Contractor to receive a response; with respect to the second category 3,586 drawing submittals took more than 25 days for the Contractor to receive a response; and with respect to the third category 1,824 drawing submittals took more than 50 days for the Contractor to receive a response. Similarly with material submittals there were excessive delays in receiving a response from the Consultants. With respect to the first category 835 material submittals took more than 15 days for the Contractor to receive a response; with respect to the second category 482 material submittals took more than 25 days for the Contractor to receive a response; and with respect to the third category 152 material submittals took more than 50 days for the Contractor to receive a response.

The Contractor also asserts that on the balance of probabilities, the Client caused the cardinal change (the material change in the nature, scope and schedule) in the Project and the Contractor suffered the effect of the events and the cumulative impact in labour productivity and other costs.

The Contractor would have been able to complete the works timeously and within budget but for the Client caused delays and disruption and, as such, the Contractor is entitled for compensation for cumulative disruption 42,684,293 USD.

References

Case Law

Bank of Credit and Commerce International SA v. Munawar Ali, Sultana Runi Khan and Others (2001) 1 All ER 961
Bernhard's Rugby Landscapes Ltd v Stockley Park Consortium Ltd (1997) 82 BLR 39 at 72

Dillingham Construction Pty. Ltd. and Others v Downs (1972) 13 BLR 97

Dimskal Shipping Co SA v International Transport Workers' Federation, The Evia Luck (1991) 4 All ER 871

Goldsmith v Rodger (1962) 2 Lloyd's Rep 249

Hedley Byrne & Co Ltd v Heller & Partners Ltd (1964) AC 465

Heilbut Symons & Co v Buckleton (1913) AC 30

J. Crosby & Sons Ltd v Portland Urban District Council (1967) 5 B.L.R. 121

London Borough of Merton v Stanley Hugh Leach (1985) 32 BLR 68

Manifest Shipping Co. Ltd. v Uni-Polaris Shipping Co. Ltd.: 'The Star Sea' (2001) 1 All ER 743

Pan Atlantic Insurance Co. Ltd v Pine Top Insurance Co. Ltd (1994) 3 All ER 581

Pao On v Lau Yiu Long (1980) AC 614

Printing and Numeric Registering v Sampson (1875) LR 19 EQ 462

Ratcliffe v Evans (1892) 2QB 524

Ray v Sempers (1974) AC 370

Servidone Construction Corporation v the United States 931 F.2d 860. April 24, 1991

With v O'Flangan (1936) Ch 575

Index

A. D. Haidar, *Global Claims in Construction*,
DOI: 10.1007/978-0-85729-730-3, © Springer-Verlag London Limited 2011